PERFORMANCE INDICATORS FOR WASTEWATER SERVICES

OPERATIONS & MAINTENANCE SPECIALIST GROUP

Chairman: Francisco Cubillo
Canal de Isabel II
Calle Santa Engracia 125
28003 Madrid - Spain
Tel.: +34 914 45 1000
Fax: + 34 914 47 9393
E-mail: fcubillo@cyii.es

International Water Association

Secretary: Enrique Cabrera Jr.
Instituto Tecnológico del Agua
Universidad Politécnica de Valencia
Apdo. 22012
46071 Valencia - Spain
Tel: +34 96 387 98 98
Fax: +34 96 387 98 99
E-mail: qcabrera@ita.upv.es

PERFORMANCE INDICATORS FOR WASTEWATER SERVICES

Rafaela Matos, Adriana Cardoso, Richard Ashley,
Patrícia Duarte, Alejo Molinari, Andreas Schulz

IWA Publishing

Published by **IWA Publishing**
Alliance House
12 Caxton Street
London SW1H 0QS, UK
Telephone: +44 (0)20 7654 5500
Fax: +44 (0)20 654 5555
Email: publications@iwap.co.uk
Web: www.iwapublishing.com

First published 2003
Reprinted 2010, 2012

British Library Cataloguing in Publication Data
A CIP catalogue record for this book is available from the British Library

Library of Congress Cataloging- in-Publication Data
A catalog record for this book is available from the Library of Congress

ISBN: 1900222906
ISBN13: 9781900222907

Printed in Great Britain by Bell & Bain Ltd.,Glasgow

FSC www.fsc.org MIX Paper from responsible sources FSC® C007785

CONTENTS

The CD content that accompanied this book can now be found at
http://www.sigmalite.com/iwamanual/

LIST OF FIGURES

LIST OF TABLES

FOREWORD

Peter Stahre

Vice Director of Malmo Water and Wastewater Works, Sweden
President of the Scandinavian 6-cities PI-group

Over the last decade there has been an increasing demand on our water and wastewater utility companies to run their businesses more efficiently and effectively. Today there is no generally accepted system for consistent comparisons of the performance of water and wastewater services. IWA has taken the leading role in elaborating a reference language for such comparisons. The definitions of performance indicators presented in this manual can together with IWA´s Performance Indicators for Water Supply Services, be looked upon as the dictionary of this reference language. The indicators describe the characteristics and the performance of individual features of utility systems. To serve its purpose each indicator must not only be very precisely defined, but also be understood and generally accepted internationally.

The numeric value of an individual performance indicator is of very little value until it is put into its context. This is for example, done by looking at how a specific indicator is developing over time or by comparing the same indicator for different utility systems. These types of comparisons are referred to as metric benchmarking. It must be emphasised here that metric benchmarking initiatives do not include the whole set of indicators presented in this manual. For each application one must select the indicators to be included, depending upon the specific purpose of the comparison. It is easy to fall into the trap of including too many indicators. As is pointed out in the manual, a PI system should preferably be implemented as a "step by step" procedure. Good advice is to "start small".

Metric benchmarking is a quantitative comparative analysis that enables a utility to follow the long-term performance of their systems. Benchmarking has proved to be a powerful management tool in the utility's strategic planning, which is, among other purposes, used for controlling the fulfilment of pre-set targets. Metric benchmarking is also used for comparisons of the performance with the performance of other utility companies. One important outcome of metric benchmarking is to identify areas of good performance as well as areas where there is a need for improvements. However, metric benchmarking does not give any explanation to why there are differences in system performance.

To explore the identified differences in system performance among different utilities it is necessary to carry out a more detailed analysis. Such an analysis is, for practical reasons, often restricted to comprise only the "worst" and the "best in class" in the metric benchmarking comparison. As a base for this in-depth analysis a selection of more detailed performance indicators and other relevant information is made. This procedure is referred to as process benchmarking. A characteristic feature of process benchmarking is that it is focusing on selected processes and not on the business as a whole. The aim with process benchmarking is to improve the performance by "learning from others".

The most advanced application of performance indicators is to rate the overall efficiency and effectiveness of the utility in quantifiable terms – so called performance assessment. In performance assessment, performance indicators are used together with other types of information (system characteristics, environmental factors etc). The relationships between the different components often form a very complex puzzle.

It is emphasised in the manual that an active involvement from the top managers and a commitment of representatives of the departments/sections involved in the data collection is critical for a successful implementation of benchmarking in an organisation. It is important that the use of performance indicators permeates the different levels of the organisation. This is especially true for the economic indicators. With a combination of a top-down and a bottom-up approach, performance benchmarking becomes a very powerful tool in improving water and wastewater utility services.

PREFACE

At the outset of the project to develop this manual, we understood that it would not be an easy task, despite the precedents and useful information already available in the water supply manual.

As with most of our work, we also expected that it would require a lot of dedication and use of personal time. In both of these perceptions we were proved to be correct.

Over the past two years a lot of time was spent under the flight path into Lisbon airport, which is where the LNEC office is located, trying to tease commonly acceptable sense out of Portuguese-English and in reconciling our various perspectives into a form we believe now has some international validity. The more junior members of the group put in the longest hours, and forgot the meaning of weekends, even when taken to work on the manual, at the final end, in the beautiful Algarve.

The range of our personal and cultural backgrounds allowed us to share a diverse view of the world's wastewater services, together with very different starting points of view. For each of us we had to challenge the paradigms, or boxes, in which we thought, so that what we hope is a truly widely applicable product, could emerge.

Yet we are still not fully satisfied with the product. Perhaps if there were a PI system on which to assess the effectiveness of manuals we might have some means of evaluating the quality of our output. We are sure, however, that this manual is just the beginning in the process of development of a worldwide process for assessment of the performance of wastewater services and service providers.

We each wish to continue to be involved in the applications of the manual and to further help it evolve along this path. If you are inspired to use it then we will be happy that we have achieved our goal and would be pleased to work with you in your endeavours.

And finally, the journey is always more important than the end product and in our journey we gained in enthusiasm, respect and friendship.

Let us hope that you take this journey and in so doing enrich your experience, challenge and broaden your thinking and pass further along the road toward truly sustainable wastewater service provision.

The authors

ACKNOWLEDGMENTS

As leader of the IWA Task Force on "Performance indicators for wastewater services" I wish to express my appreciation to all those who made this work possible.

I must first acknowledge my co-authors, Richard Ashley, Alejo Molinari, Andreas Schulz, Adriana Cardoso and Patrícia Duarte. It has been a sincere privilege to work with such an excellent team of dedicated people. Their support, encouragement and hard work were a constant incentive to pursue our commitment. Special thanks are due to Richard Ashley who additionally had the dedicated work of editorial revision. A word of appreciation is also due to Renato Parena, Chairman of the IWA "Statistics and Economics Specialist Group" for his kind and generous assistance regarding financial indicators and associated matters. Enrique Cabrera and his team, from Instituto Tecnológico del Agua in Valencia, deserve our gratitude as well for providing the software SIGMA Lite WW.

Special acknowledgments are also due to Francisco Cubillo, Chairman of the IWA Operations and Maintenance Specialist Group for his early confidence and continuous support.

The core team had several meetings during the project development. It was clearly an interesting and interactive learning experience. An enormous added value to our work came from so many people that it is not sensible to name them all.

We have received advice, comments, and suggestions from industry discussion, workshops, and widespread web discussions. The people involved represent a wide range of universities, research institutions, government agencies, wastewater undertakings and professions. To all of them the authors express their gratitude.

The continuous encouragement, help and assistance of the water supply PI team represented a major contribution. Special thanks are due to Helena Alegre for her permanent availability and friendship.

Finally I particularly wish to acknowledge my Institution, the National Civil Engineering Laboratory (LNEC), in Lisbon (Portugal), for encouraging this project and for the essential financial support during the two years project.

Rafaela Matos

Leader of the IWA Task Force on "Performance Indicators for Wastewater Services"

The CD content that accompanied this book can now be found at
http://www.sigmalite.com/iwamanual/

MEASUREMENT UNITS AND SYMBOLS

%	percentage
km	kilometre
km^2	square kilometre
kW	kilowatt
kWh	kilowatt x hour
m	metre
m^2	square metre
m^3	cubic metre
mm	millimetre
^oC	degree Celsius
US$	US dollars

ABBREVIATIONS

BMP	best management practice
CI	context information
DS	dry solids
EBIT	earnings before interests and taxes
EBT	earnings before taxes
Inhab.	Inhabitants
NGO	Non governmental organisation
No.	number
OI	operating income
p.e.	*population equivalent*
PI	performance indicator
SE	*sewer system* or *sewerage*
SUDS	sustainable urban drainage systems
WWT	wastewater treatment (includes WWTP and *on-site systems*)
WWTP	*wastewater treatment plant*

OTHER CONVENTIONS

Terms written in the format *text* are defined in Appendix 1 - Glossary.

1. INTRODUCTION

1.1 Background

Increasing diversity in the provision of services for water supply and wastewater disposal, together with greater accountability for the quality of those services, necessitate scrutiny using assessment systems that are consistent, transparent and auditable. Whilst a number of countries worldwide have well defined and regulated systems for evaluating the performance of water service providers, this is by no means global. It is within this context that the International Water Association (IWA) is developing systems of performance indicators (PIs) for water supply and wastewater services. In July 2000 a task force from the Operations and Maintenance Specialist Group of IWA produced the IWA Manual of Best Practice "Performance Indicators for Water Supply Services" (Alegre *et al.*, 2000). The original document is available from IWA Publishing, however, on-going testing with some 70 organisations worldwide has resulted in a number of changes to the approach recommended and the PI system. The lessons from the testing of the water supply PI system have been used to inform the development of this manual, an equivalent document which deals with Performance Indicators for Wastewater Services. The approach to the specification, selection and usage of the PIs is generic to both water supply and wastewater services. Although the particular PIs recommended for use for performance assessment of each of the two systems are not the same. The IWA PI systems are intended to provide objective and comprehensive management tools for undertakings and other *stakeholders* involved in any aspect of water and wastewater service provision. The PIs have been developed independently from the level of economic development, or type of Institutional system in which the undertaking operates, and allow globally diverse economic, demographic, cultural and climatic characteristics to be acknowledged. The PIs cover a wide range of activities comprising: management, personnel, financial, physical, operational, environmental and quality of service aspects. The PIs for wastewater systems and associated procedures presented in this manual are expected to undergo further development through wide-scale pilot testing.

1.2 Why use performance indicators

Irrespective as to whether it is a private company or a public municipal service supplier, any undertaking needs to strive for high degrees of efficiency and effectiveness to achieve its management goals. In addition, other stakeholders, such as regulators or customers require assurance that the undertaking is performing appropriately. With increasing diversity in the way in which water services are provided, a recent state-of-the-art review, covering more than 16 countries, concluded that some form of standardised performance assessment has become essential (Merkel, 2002).

Performance Indicators (PIs) may be considered as providing key information needed to define the efficiency and effectiveness of the delivery of services by an undertaking (Deb & Cesario, 1997). Efficiency is the extent to which the resources of an undertaking are utilised to provide the service, for example, maximising service delivery for the minimum use of available (possibly natural) resources. Effectiveness is the extent to which declared or imposed, objectives, such as levels of service, (specifically and realistically defined) are achieved. PIs may also be considered as providing information for 'metric' benchmarks – quantitative comparative assessment of performance (Larsson *et al.*, 2002). The actual comparison of performance between similar service provision is undertaken via 'process' benchmarking – examining business processes, comparing the activities of different organisations and seeking to identify best practices. A performance indicator may thus be used as a quantitative (or in some cases qualitative) measure of a particular aspect of an undertaking's performance or standard of service. PIs may be used to compare performance historically, or against some pre-defined target (Alegre *et al.*, 2002a; Matos *et al.*, 2002b). PIs may be used by a wide range of stakeholders in evaluating the performance of the undertaking, including internal evaluation within the undertaking itself.

As an example of PI use, in 1995 the six-Cities Group in Scandinavian countries (Copenhagen, Gothenburg, Helsinki, Malmo, Oslo, and Stockholm) initiated the development of a coherent performance benchmarking system for water and wastewater services (Stahre & Adamsson, 2002). They have developed a set of performance indicators that may be considered as a standardised reference language necessary for making consistent system comparisons. Annual benchmarking exercises are carried out and the results presented in two forms, in summary, to the Managing Directors of the undertakings and in more detail to the operational staff. Comparative performance between the six undertakings and trends looking back five years are also highlighted. The metric benchmarking processes have also been supplemented by process benchmarking whereby differences in performance between the organisations are investigated.

Precise definition of the PIs to be used by each *stakeholders* and within stakeholder subgroups is not straightforward. The use of indicators in general is problematic, and has received a lot of attention with regard to attempts to define sustainability (e.g. Ashley *et al.*, 1999). A major difficulty is where the PIs are used in a judgemental way. One example of this is the way in which the Regulators assess the economic performance of the water undertakings in England and Wales. The performance objectives and targets are produced jointly by the Water Companies, the Government and the Water industry economic Regulator, the Director General of Water Services and Office of Water Services (OFWAT). Targets are set in association with other bodies, NGOs, consumer representatives and the Environment Agency. In the UK sector the three critical areas of performance judged by the regulatory bodies are value for money, service delivery and environmental and social impacts. The UK regulatory approach is essentially judgemental and comparative to assess the performance both between companies and against pre-determined targets. The actual PIs in each specific application that are used reflect this perspective (Ashley & Hopkinson, 2002).

Recently, a number of regulatory approaches have addressed asset serviceability. This differs from the traditional levels of service concepts, by being concerned with the longer-term service provided by capital assets. This necessitates capital maintenance in order that assets may continue to deliver services to customers (Reynolds, 2000). There is a need for this approach where current customers are receiving a satisfactory service (as defined by current performance in terms of levels of service), but where there appears to be no incentive for longer-term investments by the undertaking for future customers. In the UK, the approach to PIs to assess asset serviceability is now risk based (Ewan Associates & Mott MacDonald, 2001). In this, asset serviceability is defined as the ability of an asset to deliver a defined service to customers and safeguard the environment. There are four levels of PI. At level 1, the base data are used to reflect the state of the undertaking's assets and the impact on customers and the environment. The higher level indicators require further analysis or combinations of the base PIs in order to assess trends in serviceability, i.e. as an indication of a potential declining trend in serviceability. In the IWA PI manuals the indicators provide the base information equivalent to the UK Level 1 serviceability indicators.

Whilst the IWA PIs are intended to be comprehensive (covering all forms and contexts for water service provision), they are meant to be informative rather than judgmental. It is expected that much of their usage will be within organisations for self-improvement with regard to specific aspects of the undertaking's activity and of the system behaviour. This may also be informed by self-comparison (benchmarking) with other similar water service providers operating with similar constraints and contexts.

Given that as yet it is not possible to define what may or may not be a sustainable system, moves toward greater sustainability in water services can only be achieved using criteria and indicators, such as the PIs described here, that inform judgement as to the best way of providing the service (Ashley *et al.*, 2002). Hence the approach being used is a key component in building systems that are likely to be as sustainable as current knowledge allows. In this context sustainability means both the traditional definition (including social-environmental-economic criteria) and the undertaking definition, which is that the service provider remains in business (Foxon *et al.*, 2001).

1.3 Usage of performance indicators

Performance indicators can be of use to a wide range of *stakeholders* (e.g. Audit Commission, 2000). However, it is envisaged that the PIs presented in the IWA manuals will be utilised primarily by the service providers. The following outline the reasons why various stakeholders may use the PIs.

For *water supply* and *wastewater undertakings*:

- To assist with strategic and structured planning;
- To facilitate better quality and more timely responses from managers;
- To allow for simpler and more structured monitoring of the effects of management decisions;
- To provide key information to support a more pro-active approach to management, rather than reacting to apparent system or organisational malfunctions;
- To highlight strengths and weaknesses of individual departments, or regional sectors of the organisation, demonstrating where there may be a need to improve productivity, procedures and routines;
- To assist with the implementation of a Total Quality Management regime, as a way of improving all-round quality and efficiency throughout the organisation;
- To facilitate the implementation of benchmarking routines, both internally, for comparing the performance at different locations or for sub-systems, and externally, for comparison with other similar undertakings, thus providing key information to promote future performance improvements;
- To provide sound scientific, technical, financial and personnel information for auditing the workings of the undertaking and to help to predict the likely effects of any recommendations made as a result of an audit.

For national or regional *policy-making bodies* and *Regulatory Agencies*:

- To provide a consistent framework for comparing the performance of wastewater undertakings and identifying areas of activity where improvement is required;
- To support the formulation of policies for the sector, ideally within a perspective for the integrated management of water resources, including environmental protection, resource allocations, investments, and the development of new regulatory tools.
- To provide key monitoring information to help safeguard customer interests where service is provided by a monopoly, particularly to monitor compliance with contracted service level goals.
- To provide evidence of adequate environmental performance by comparison with pre-defined standards.

For *financing bodies*:

- To assist in assessing investment priorities, risks, project selection, realisation and audit.

For *consumers* and *their representative bodies*:

- To provide wide-ranging standardised information about the performance of the undertaking in a way as transparent as possible.

For *Quality Certifying organizations*:

- To provide key information for quality assurance purposes.

For *Auditors and Economic Regulators:*

- Use as part of a consistent framework for auditing asset value, condition, and financial and related performance.

For *multi-lateral organisations*:

- To provide a tool for identifying the main differences in performance of services between regions, and hence to inform future strategies for improvement and the allocation of resources.

The wastewater PIs may be used by these various organisations to evaluate performance historically, regionally or nationally, against previous time periods or comparable organisations. Trends in PIs with time may show historically improving or deteriorating performance in time for remedial measures to be taken before major problems occur in service delivery. Where new systems are being implemented, PIs may show whether these are achieving effective and efficient performance compared with other organisations. This is particularly important in developing countries. PIs require a lot of data; hence they can also provide the incentive for good monitoring, data recording and processing, and help decision makers focus scarce resource allocations into the key areas where data needs to be collected.

The main forms of using performance indicators are (adapted from Alegre, 2002b):

- *Exclusively within the undertaking* – managers may use PIs to monitor the evolution of performance, comparing the results achieved at a given time with those achieved in previous periods. Another form of use is comparing the results actually achieved with preset targets or with reference values that have been published from other organizations. For the former, the undertaking may define internal indicators, whereas for the latter, the undertaking has to adopt standardized definitions to avoid non-commensurate comparisons.
- *In the framework of benchmarking initiatives* – the internal use of PIs can be complemented with metric benchmarking initiatives; these can be held within a group of undertakings that can agree among them a set of indicators to be adopted by all the group members, based on a common set of definitions. The analysis, interpretation and eventual publication of the results must always take into account these definitions.
- *As part of a regulatory framework* – the worldwide growth of private-sector participation in the management of water supply and wastewater systems requires added vigilance from economic regulators. For example, the adoption of PI systems may be used as part of a surrogate for true commercial competition among water service providers. Where the performance of an undertaking is being evaluated against another the use of standardised definitions for PIs is essential.
- *As part of contractual agreements* – traditionally, contract agreements between the municipalities and private operators are very detailed in terms of the financial arrangements, but may be less specific in terms of the quality of service to be provided to the customer. The use of PIs provides a means whereby specified and scheduled targets may be defined in an objective and auditable way, protecting the customers' interests. Whilst the use of customised PIs for this is feasible and sometimes necessary, however, the use of standardised PIs provides a more objective, independently accredited and comparable foundation on which to base agreements.
- *As part of Quality Certification Systems* – monitoring of the undertaking's internal processes to attain ISO 9000 standards or other certification can be assisted by using PIs. Utilisation of standardised PIs provides a more robust foundation for certification than self-selected PIs.

- *To provide public domain statistical reports* – PIs may provide the baseline statistics for citizen information, or to be made available collectively by organisations such as Waterworks Associations, regionally or internationally (e.g. World Health Organisation, Asian Development Bank, World Bank). Traditionally, this type of performance monitoring has been limited to statistical data and economic indicators, but recently other types of indicators are also being presented, dealing for example with environment. Unfortunately information presented in these reports is usually inconsistent, and varies considerably between different undertakings (e.g. Hopkinson & Whitaker, 1999). It may also be of doubtful value due to a lack of precise definitions for the indicators. The use of consistent and standardised PIs is therefore very important in these applications.

1.4 Performance indicators as a component of benchmarking

The implementation of objective-oriented management procedures is an indispensable step for the success of most companies (PBM SIG, 2001). Whether private or public sector, the performance of water and wastewater undertakings is assessed typically in relation to better performing companies. The objective-oriented approach requires: the establishment of clear objectives to be achieved within given deadlines; the comparison between targets and results; and the correction of the causes for the deviations between target and results, so that the performance of the undertaking may improve. Performance indicators can be a source of key information in this context as they provide clear and quantified data that may be used for comparative assessments. Some companies have realized that if they compare themselves with the best performers, and correctly identify the reasons for any discrepancies, they can often improve their performance significantly. This is how benchmarking appeared and was successfully utilised in many industrial sectors. Such process benchmarking is becoming more popular in the water sector (e.g. Larsson *et al.*, 2002), and it is clear that any comparison between different undertakings requires the use of standardised PIs. A commonly consulted worldwide metric benchmarking methodology, is the Benchmarking Start Up Kit developed by the World Bank (World Bank, 2002).

In the UK, the economic regulator uses the relative efficiency between the water service companies as a means for determining future asset management investment policy (Ashley & Hopkinson, 2002). In this case the benchmark is an aggregate measure across the more efficient performers amongst the private companies, and taking into account a limited number of indicators. For the next UK water price review, the process whereby the future price of water services will be determined will be based on indicators and procedures defined in The Common Framework (UKWIR, 2002). A 2001 review of International Benchmarking, the Water Research centre (WRc) in the UK stated that benchmarking was useful, but simply a tool that *must form part of a wider performance improvement strategy* if it was to help to enhance performance. Metric benchmarking, which is used to compare data via PIs, may not identify or exchange information about best practices. Whereas *process* benchmarking, does examine business and other organisational processes, comparing activities of different organisations, so identifying best practice. Hence PIs are an essential component on which to base process benchmarking exercises designed to enhance performance by inter and cross-comparisons with others.

It is not a straightforward process to assess how efficient and effective an undertaking is even if formalised processes, such as project management or quality management systems are used. Performance can be assessed by comparison with other similar service providers. PIs may be used as the first stage in this comparative evaluation. PIs may be compared with historical information, other departments or undertakings. A formalised PI-system is needed that takes into account all influences on the performance for effective comparison, hence the need for the IWA-PI-system presented here. The outcome from such a comparison is the first step in the overall evaluation process (Schulz & Schön, 2001).

The next step should be wider *stakeholder* participation in evaluation of performance. As part of this process the undertaking should investigate 'best practice' as provided by other service providers and make this information widely available. If the undertaking wants to improve performance engagement in a benchmarking project may be effective. Benchmarking in this context may be defined as *a systematic continuous process to identify and adopt successful instruments, methods and processes of other organisations* (Schulz & Schön, 2001). It is most effectively carried out where a dialogue can be established with other similar undertakings (ideally with a better performance) by comparing (anonymous) results from a wide range of (similar) service providers. By this means different operational contextual conditions can be better taken into account and the ideal *best practice* of the undertaking identified.

The result of a Benchmarking project is the better definition of the areas of performance of the undertaking that should be improved. Hence the success of implementing a PI system depends on the effective use of the results as part of a decision-making tool. The objective is to interpret the results, i.e. to decide whether a given result is satisfactory or not, and if not to establish a diagnosis and to identify what improvement measures need to take place.

The process of result interpretation requires comparison with a reference value or a reference threshold, taking into account the context information, and analysing a group of related indicators rather than single indicators individually.

The sources of benchmark reference values are of five main types:

- Equivalent results from earlier time periods;
- Targets established in strategic plans or otherwise self-imposed;
- Targets established within legal, contractual or regulatory frameworks;
- Results from other operational units of the same undertaking (e.g. valid for holdings managing more than one undertaking or for undertakings with more than one system or operational unit);
- Results from other undertakings using similar metric benchmarking initiatives;
- Reference values that are publicly available.

The benchmark selected will depend on the objective of the PI assessment and the context within which it is being carried out.

1.5 International Standardisation of water supply and wastewater services

A clear signal of the worldwide relevance of a comprehensive framework of performance indicators in the scope of water supply and wastewater services provision is the recent developments of International standards as part of the work of the Technical Committee ISO/TC 224 – *Service activities relating to drinking water supply and wastewater systems – quality criteria of service and performance indicators.* The scope of ISO /TC 224 is the standardization of a framework for the definition and measurement of service activities relating to drinking water supply and wastewater systems. The standardization includes: the definition of the characteristics of the elements of the service according to customer expectation, a list of requirements to fulfil for the management of drinking water supply and wastewater systems, service quality criteria and related system of performance indicators, without setting any target values or thresholds. Excluded from this scope are the design and the construction of water supply and wastewater systems, maintenance techniques, the limits of acceptability for drinking water quality and wastewater discharged in the receiving body and analytical methods (ISO/TC 224, 2003).

ISO/TC 224 has established the following four Working Groups: WG1 – *Terminology*; WG2 – *Service to the consumers*; WG3 – *Management of drinking water systems;* and WG4 – *Management of wastewater systems.* Standards are expected to be published by 2006.

ISO/TC 224 has twenty participating members (Argentina, Austria, Belgium, Canada, France, Germany, Japan, Republic of Korea, Malaysia, Mexico, Morocco, Netherlands, Norway, Portugal, Russian Federation, Slovakia, South Africa, Spain, United Kingdom, United States of America) and twenty observer members (Australia, Bulgaria, Colombia, Cuba, Czech Republic, Denmark, Ecuador, Finland, Hungary, Ireland, Israel, Italy, New Zealand, Nigeria, Poland, Sweden, Switzerland, Turkey, Yugoslavia, Zimbabwe). Thus, most industrialised nations and a significant number of developing countries are represented. ISO/TC 224 also has liaisons with a world wide range of international organisations such as Consumers International, Eureau, Norm APME (European Office of Crafts, trades and small and medium-sized enterprises for standardisation), World Bank, World Health Organisation, International Water Association, CEN/TC 164 – *Water Supply* and CEN/TC 165 – *Wastewater engineering*.

The IWA PI manuals will contribute directly to the work of the Technical Committee. The piloting of the water supply manual, with subsequent refinement of the PIs and procedures, together with the production of the wastewater PI manual, and subsequent pilot testing will provide key information that will help in defining the standards and procedures that the ISO will eventually stipulate. As a Corporate member of the Technical Committee, IWA can readily provide the results from the pilot testing and recommendations for the refinement of the PI procedures. In addition a number of the authors of the two manuals are also representative country members of the Technical Committee and can contribute directly to future development of the standardisation framework.

2. DEVELOPMENT OF A MANUAL FOR WASTEWATER SERVICE PIS FROM THE WATER SUPPLY MANUAL

2.1 The IWA manual of PIs for water supply

The PIs manual for water services was published in 2000, as described in Section 1.1. Since then the manual and PI system have undergone detailed evaluation via pilot testing by some seventy users. The testing has resulted in a number of enhancements to the approach recommended and the PI system.

From the reviews and pilot studies of the water service PI manual, the main positive characteristics of the IWA-PI system are listed below (Alegre *et al.*, 2002b):

- comprehensiveness (it is perceived to cover all the main activities of a water supply undertaking);
- the structure adopted, including detailed definitions of all variables required to assess the indicators and a clear identification of the processing rule for each indicator; in addition, the tables of variables includes the identification of all of the indicators each variable is used for, directly or indirectly through other variables;
- ease of selection of subsets of relevant PIs applicable to specific cases;
- standardisation of PI terminology;
- availability of the software SIGMA;
- potential for future comparisons with other undertakings;
- possibility of customisation by splitting existing PIs in terms of geographical area, time period or components within the same structure, without losing the ability to aggregate and standardise values;
- possibility of adding new PIs and adopting the same overall structure.

The main concern expressed by users was the apparent complexity for first-time application and the impression that all 133 water supply indicators and corresponding variables had to be used. In fact, undertakings are not expected to use the whole set at a time, and the PI system (and the software SIGMA) is tailored to allow for an easy selection of the indicators that a given undertaking considers relevant for specific applications. The corresponding variables required for the subset selected are easily (or even automatically, in SIGMA Lite) identified. As a result of the trials, the project team have produced a new layout for the second version of the Water Supply Manual. The generic approach has not been significantly modified, but guidance will be clearer on implementation and a wide range of examples will be included.

In the original PI manual for water services, recommendations for three levels of importance and suggestions for phased application of sub-sets of the large number of PIs, were suggested. This division was considered important at the time to allow a phased implementation of subsets of the full suite of PIs, avoiding a potentially too demanding task for the undertakings from inception, as PIs requiring greater detail (and more data) could be implemented sequentially (Alegre *et al.*, 2000). Pilot testing revealed that it was not possible to define three generic subsets from the full suite of PIs that were applicable and of fixed relative importance for all situations. Thus the various users of the manual perceived different subsets of the PIs to be of differing relative importance, depending upon their individual objectives, context of application and availability of data. In order to deal with these difficulties, manual users were recommended to assign their own relative levels of significance to individual PIs, targeting first those perceived to be the most significant.

2.2 Lessons from the water supply PI pilot testing

Following the three years since publication of the water services PI manual, and pilot testing, it is apparent that the general approach and format used has been appropriate and widely welcomed. Hence the wastewater services PI manual has maintained the generic principles and approach set out in the original water services manual, in that the PIs have been defined in a similar way and their usage incorporated into a similar system framework. The improvements to the water supply PI manual methodology, following the three year evaluation, have enabled the wastewater service performance manual to be developed to be more directly responsive to user needs and perspectives.

It is constructive to consider the primary differences between the water supply and wastewater PI requirements and perspectives, adapted from Duarte *et al.*, 2003:

- Water supply services are directed principally to the consumers situated at the end of the system, whilst wastewater services aim primarily to assist the population to maintain health and welfare by removing pathogens and preventing *flooding*. Also to assist commercial customers to dispose of undesirable effluents. This difference is significant for the assessment of quality of service.
- Environmental impacts of water supply systems tend to be less significant than wastewater systems. In the former the most important performance measures may relate to abstractions, where these involve for example, large impoundments. In addition the overall efficiency of use of the water resources will be significant. For wastewater, the undertaking performance may control the physical impact of flood related flows and pollution caused to environmental systems where *appropriate treatment* is not utilised.
- Wastewater systems have potentially a greater environmental impact than water supply services due to flood risk and environmental pollution. Wastewater also has a wider variability in terms of quality due to the virtually infinite variety of contaminants that may be introduced via the wastewater system. Hence the wastewater manual has more PIs dealing with treatment processes than does the water supply manual.
- Water supply services are primarily to serve consumers, hence the main effects of a poorly managed water supply service will impact principally on these consumers, many of whom will be customers. The exceptions here may be where abstraction is badly managed or water main bursts are endemic, in which case a much wider range of *stakeholders* may be affected. In general, a wider range of stakeholders, not only customers, will be affected where wastewater services are badly managed. This will occur due to either *flooding* or pollution.
- Water supply systems are in general more definable than wastewater systems. With the exception of the original source, water supply systems are operated by well-controlled management systems. Once water has been abstracted, the performance of water supply systems is thus reasonably predictable. Wastewater systems are, however, essentially random processes, as the primary driver is precipitation. Even in separate sewers conveying *dry weather flows*, *infiltration* and *wrong connections* often result in highly variable behaviour. Hence it is more difficult to predict, measure and control the behaviour of wastewater systems compared with water systems. This results in the necessity to collect (or predict using models) much more data for wastewater system performance than for water systems.

The water supply and wastewater PIs share a common structure, are sub-divided into (almost) the same classification groupings, and have many indicators and variables in common. However, the important differences in core characteristics and service requirements between water supply and wastewater systems described above necessitate a number of differences between the two PI systems. The principal differences between the approach used in the wastewater PI system and manual and what is given in the original water supply manual (Alegre *et al.*, 2000) are summarised below.

- Added flexibility for the reporting time period. In the original manual this was fixed at one year. In the wastewater manual and software the user has freedom to specify the time period over which the data have been collected. This is defined here as the *assessment period*. It is important that the users of the manual and software are aware that there may be a number of different time periods used within the data sets. Hence care must be taken when making any comparisons or when evaluating performance.
- There are two types of PI that are different from those presented in the original water supply manual. One type has been developed in conjunction with the review of the water supply manual and shares some commonality with the revisions required to that manual. Most of these PIs are in the economic/financial category. For example, the definition of a new set of PIs dealing separately with annual running costs for different aspects of each part of the technical service provision per unit of activity. The second type includes PIs that are exclusive to the wastewater service industry and of no relevance to water supply services. Of these, there are for example, a much larger number of environment related PIs. An example of this relates to the *population equivalent* that is served by wastewater treatment plants that comply with discharge consents.
- More detailed guidance on the PI selection and implementation strategy is provided in the wastewater services manual.
- As part of the above, a procedure is provided for assessment and prioritisation of PIs in terms of their relative significance for the objective or context of the application in the wastewater services manual.
- Infrastructure *rehabilitation* concepts and definitions were not clear enough in the water supply manual for a number of users. More detailed definitions have been provided in the wastewater manual to clarify definitions and usage of the PIs related to rehabilitation, *renovation*, *replacement*, *reinforcement* and *renewal*.
- As definitions and understanding of terminology related to wastewater systems varies around the world, it was deemed prudent to include an extensive glossary of terms in the wastewater manual to avoid confusion and misconceptions.

3. ABOUT THIS MANUAL

3.1 Scope

The main purpose of this manual is to present a coherent framework of performance indicators (PIs) for wastewater services. The manual also provides guidance as to how to utilise the PIs or subsets thereof for:

- internal evaluation of performance by wastewater undertakings;
- comparisons between the performance of undertakings in the scope of metric benchmarking initiatives;
- regulatory agencies to extract requisite information for their specific needs;
- to provide data for international comparisons of the performance of wastewater services.

The current stage of development of the manual aims to include the performance indicators most relevant for more effective and efficient performance management of a wastewater undertaking. These PIs need to incorporate all the relevant aspects required to address management needs and objectives in terms of evaluating and enhancing the performance of the undertaking. Complementary indicators may be needed at departmental level, but these tend to be much more dependent on the structure of the particular organisation. Therefore, they are not specifically included here.

The manual is intended primarily for use by wastewater undertakings, but may also be used by other *stakeholders* concerned with the performance of an undertaker in providing wastewater services. Currently it is not usual for undertakers to be responsible for all parts of the wastewater system, which is shown diagrammatically in Figure 1. In particular, those parts of the system that deal only with *stormwater*, which does not ultimately enter a sewer operated by the undertaking, are excluded. This excludes stormwater handled by *source control* systems, *BMPs* or *SUDS*. However, *on-site systems* dealing with *sanitary sewage* are included if they are the responsibility of the undertaking. Where on-site sanitation is practiced at household level and is locally managed, this is also excluded apart from any services provided by the undertaking, such as the emptying of *septic tanks*. The definitions of the terms related to the wastewater system are given in the Appendix 1.

3.2 Structure of the document

The performance indicators presented in this manual may be considered as comprising a language and therefore the manual contains the vocabulary, the definitions, the system in which they are framed and the instructions for usage.

After introductory Sections 1, 2 and 3, Section 4 introduces relevant aspects regarding the wastewater PI system, including the PI and CI concepts (4.2), the wastewater undertaking context (4.3), the wastewater system description (4.4), organisational functions (4.5), financial aspects (4.6) and the structure of the PI system (4.7).

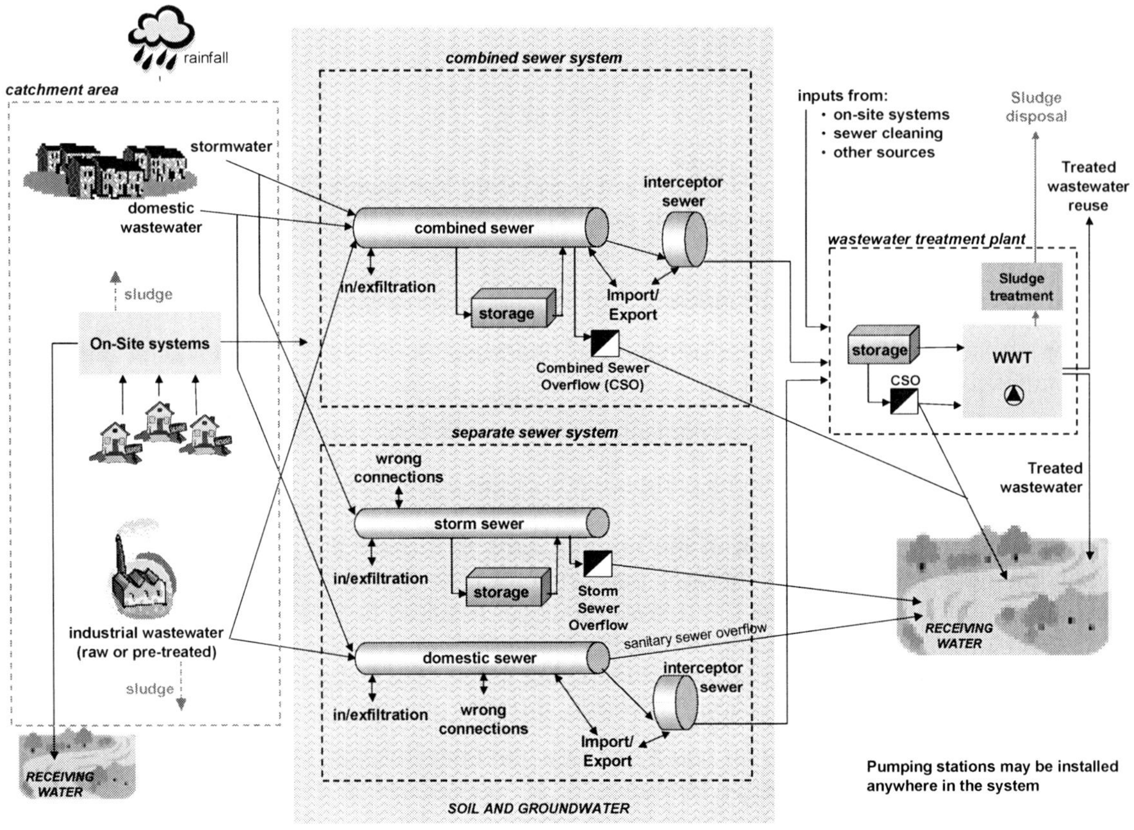

Figure 1. Wastewater system.

Section 5 proposes a phased strategy for the implementation of the IWA wastewater PI system, which includes the definition of the strategic performance assessment policy (5.2), the selection of PIs to be assessed (5.3) and the implementation of important PIs (5.4). These phases are followed by the PI assessment, reporting and result interpretation and by the definition of improvement measures, both described in (5.5) and (5.6).

Section 6 is related to data reporting concepts, including the confidence grading scheme (6.1 to 6.3) and periods for data assessment (6.4).

Section 7 contains the full listing of the performance indicators for wastewater services. The PIs are structured into six groups: environmental (7.2), personnel (7.3), physical (7.4), operational (7.5), quality of service (7.6) and economic and financial (7.7).

Section 8 deals with context information structured into: undertaking profile, system profile and region profile. The undertaking profile (8.2) outlines the framework of the organisation. The system profile (8.3) focuses mainly on the type of wastewater service provided, on the physical assets, on the technological means used and on the customers. The region profile (8.4) provides information about the demographic, economic, geographical and environmental context.

The Appendices complement the main text: Appendix 1 includes a comprehensive glossary of terms related to the wastewater system and to population; Appendix 2 provides the definitions of each of the input variables required to derive the proposed performance indicators; Appendix 3 gives a fictitious example of the relative significance of each of the PIs to a typical wastewater undertaking in a developed country and finally Appendix 4 provides guidance for the use of the SIGMA Lite WW software.

4. THE WASTEWATER PI SYSTEM

4.1 Outline

This chapter deals with the structure and nature of PI systems and how these have been applied in the manual to provide metric benchmarks for the performance of wastewater systems and service providers. It is necessary firstly to elaborate further and specifically, the earlier explanations about PI systems and how they will be utilised here (4.2). Secondly, the context within which wastewater undertakings operate and the nature of the systems they operate is outlined in terms of the nature, fluxes and demographic aspects of these systems (4.3 and 4.4). Then the various aspects of the organisational functions of wastewater undertakings are considered, together with the personnel (4.5) and financial (4.6) aspects of operation. Finally the structure of the PI system to be used to evaluate the performance of wastewater service providers is then presented (4.7).

4.2 The PI and CI concepts

Performance Indicators (PIs) are typically expressed as ratios between variables; these may be commensurate (e.g. %) or non-commensurate (e.g. $/m^3). It is important to comply with a range of requirements for the definition of the PIs. Individual PIs should be unique and collectively appropriate for representing all the relevant aspects of wastewater undertaking performance in a true and unbiased way, thus reflecting the managing activity. They are also required to be applicable to undertakings with different characteristics and stages of development. Furthermore, it is important that PIs are clearly defined, with a concise meaning and a unique interpretation for each indicator, easy to understand even by non-specialists, easily verifiable (auditable), self-explanatory and always related to well-defined areas and periods of time (Alegre *et al.*, 2000 and Matos *et al.*, 2002a). Each performance indicator should contribute to the expression of the level of actual performance achieved in a certain area and during a given period of time, allowing for a clear comparison with targeted objectives and simplifying an otherwise complex analysis.

The interpretation of performance of an undertaking cannot be assessed adequately without taking the context in which it operates into account, as well as the most relevant characteristics of the system and of the region. The IWA PIs are, as a consequence, informed by 'Context Information' (CI) considered in terms of profiles for: the undertaking; the systems managed; the geographical context in which the undertaking operates.

As described in more detail in chapter 8 the undertaking profile outlines the organisational structure for the undertaking. The system profile focuses mainly on the type of wastewater infrastructure and service provided, i.e. the physical assets, technology used and type of customer. The latter requires more detail than the other profiles because it also contains descriptive information that is helpful for the interpretation of the PIs. The region profile is essential for meaningful comparisons between undertakings as it allows for a better understanding of the demographic, economic, geographical and environmental context.

The six-Cities group described in Section 1.2, reviewed their six-years of experience of performance benchmarking and the following recommendations have been adapted from this review:

- Clear definitions of basic data and performance indicators are of utmost importance – the quality and reliability of the performance indicators is highly dependent on how well they are defined. The group had to evolve their PI definitions with experience.
- Do not collect too much information – try to keep it simple - It is tempting to want to collect too much information. After four years the numbers of PIs were revised. As a result the number of collected data items was reduced to almost half of those originally specified.

- Follow the long-term development of a selection of significant performance indicators - by following the long-term development of a selection of the most significant performance indicators it is possible to make important conclusions about the more efficient and effective management of the service provision.
- Comparison of performance indicators is a first step towards process benchmarking - the comparison of performance indicators provides a good foundation for more in-depth investigations between undertakings about differences in performance, costs, working routines etc. These investigations provide further momentum for process improvements.
- Information about performance indicators must be shared throughout the organisation - information about the use and development of PIs can stimulate staff throughout the organisation to contribute to the planning and implementation of improvements in service delivery. It is therefore important that PIs are made available at all levels in the undertaking.
- Active involvement and commitment from the most senior level is necessary - interest in working with PIs has progressively increased in the six-cities Group. The main reason for this has been very active involvement and participation from the most senior staff.

The original water supply manual specified the fundamental requirements to be taken into account when specifying and selecting PIs. These original requirements have been further endorsed by the testing of the of the water supply manual and also by other studies, such as that of the six-Cities group above. The same requirements have therefore been adopted in the wastewater manual, and are re-stated below:

- PIs should represent all the relevant aspects of the performance of a wastewater undertaking, allowing for a global representation of the system by a reduced number of performance indicators;
- PIs should be suitable for representing the relevant aspects of the performance of an undertaking in a true and unbiased way;
- A PI should be clearly definable, with a concise meaning and a unique interpretation for each indicator;
- PIs should each be mutually exclusive without overlap;
- The data needed for each PI should, as far as practicable, require measuring equipment or other data generation methods, such as computer modelling, that is affordable, avoiding the need for sophisticated and expensive equipment;
- Each PI should be auditable - especially important when the PIs are to be used by regulatory bodies that may need to check the results reported;
- PIs should be easy to understand, even by non-specialists e.g. consumers, and as wide a range of *stakeholders* as possible;
- The PIs should refer to a well-defined period of time (one year is the assessment period recommended, although in some cases other periods may be appropriate);
- PIs should refer to well-defined geographical or spatial areas;
- PIs should be applicable to undertakings with different characteristics and at various stages of development;
- The PIs should be as few in number as possible, including only those deemed absolutely essential for effective performance evaluation.

Some of the above requirements appear contradictory. For instance, the number of indicators required to provide enough information for effective performance assessment might appear excessive. However, users should be aware that there are potential dangers when using too few PIs. The evaluation of performance using too few indicators can be misleading – this is often due to the 'push-pull' effect. This occurs where improvements in one area consequentially cause deterioration in performance in another area. Hence the PIs selected must include those that are mutually influenced by one another. Historically, the environmental performance of industry has often suffered from an inadequate consideration of this effect.

4.3 The wastewater undertaking context

It is necessary to understand the context within which an undertaking has to deliver services. The context defines the mission, the constraints, interactions, needs, outputs and most importantly the financial framework within which the undertaking has to operate. The context thus defines the nature of the service and boundaries within which the service has to be provided. Typically an undertaking will wish to provide good quality services whilst simultaneously protecting the environment and being financially viable. Increasingly undertakings are also wishing to be seen as socially responsible. The context within which a wastewater undertaking has to operate is illustrated diagrammatically in Figure 2.

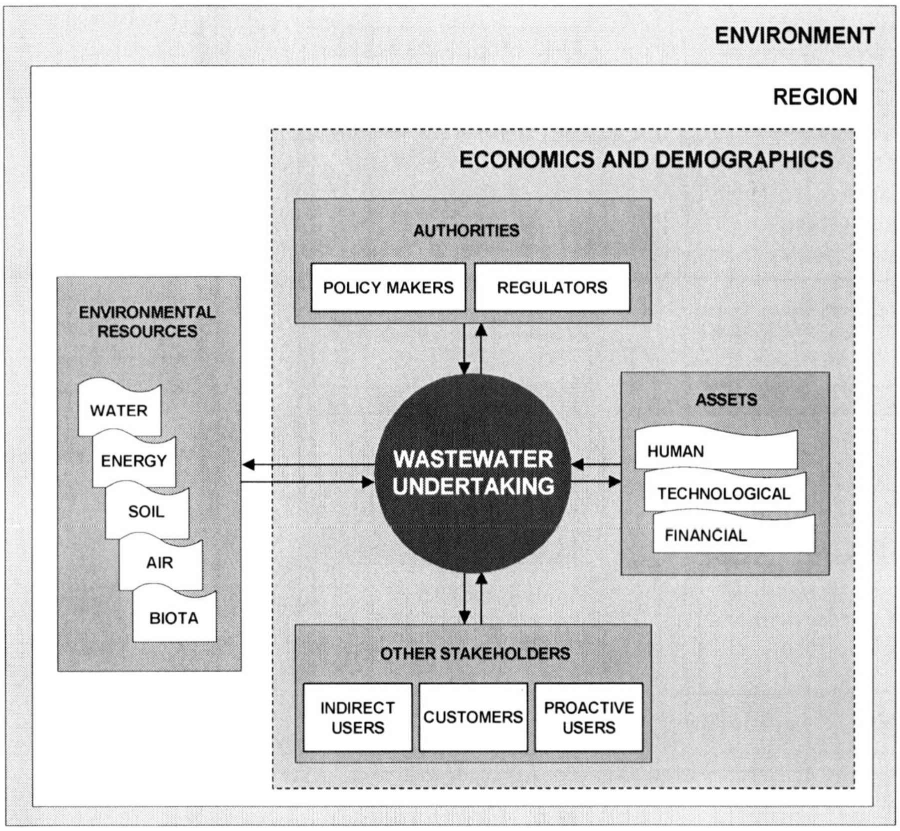

Figure 2. Wastewater undertaking context.
(adapted from Alegre *et al.*, 1997)

The *stakeholder* groups include those directly serviced by the undertaking, the customers, and also indirect users of the services who are not customers. This may include for example, recreational users of coastlines into which the undertaker discharges effluents. The proactive users include NGOs, pressure groups etc., comprised of stakeholders who may or may not be customers of the undertaking. Where the undertaking is privately owned, the *shareholders* or owners are considered here as part of the main undertaking group.

4.4 Wastewater system description

There are a number of ways in which a wastewater system may be defined. Figure 1 shows the main components and linkages from an infrastructure perspective. An alternative representation is shown in Figure 3, where flow, conveyed solids and other waterborne substances are illustrated as a simplistic wastewater balance, or flux. In this, the sum of the input components and the system losses is equal to the outputs from the system. The definitions of the terms in these Figures are given in the Appendix 1 - Glossary.

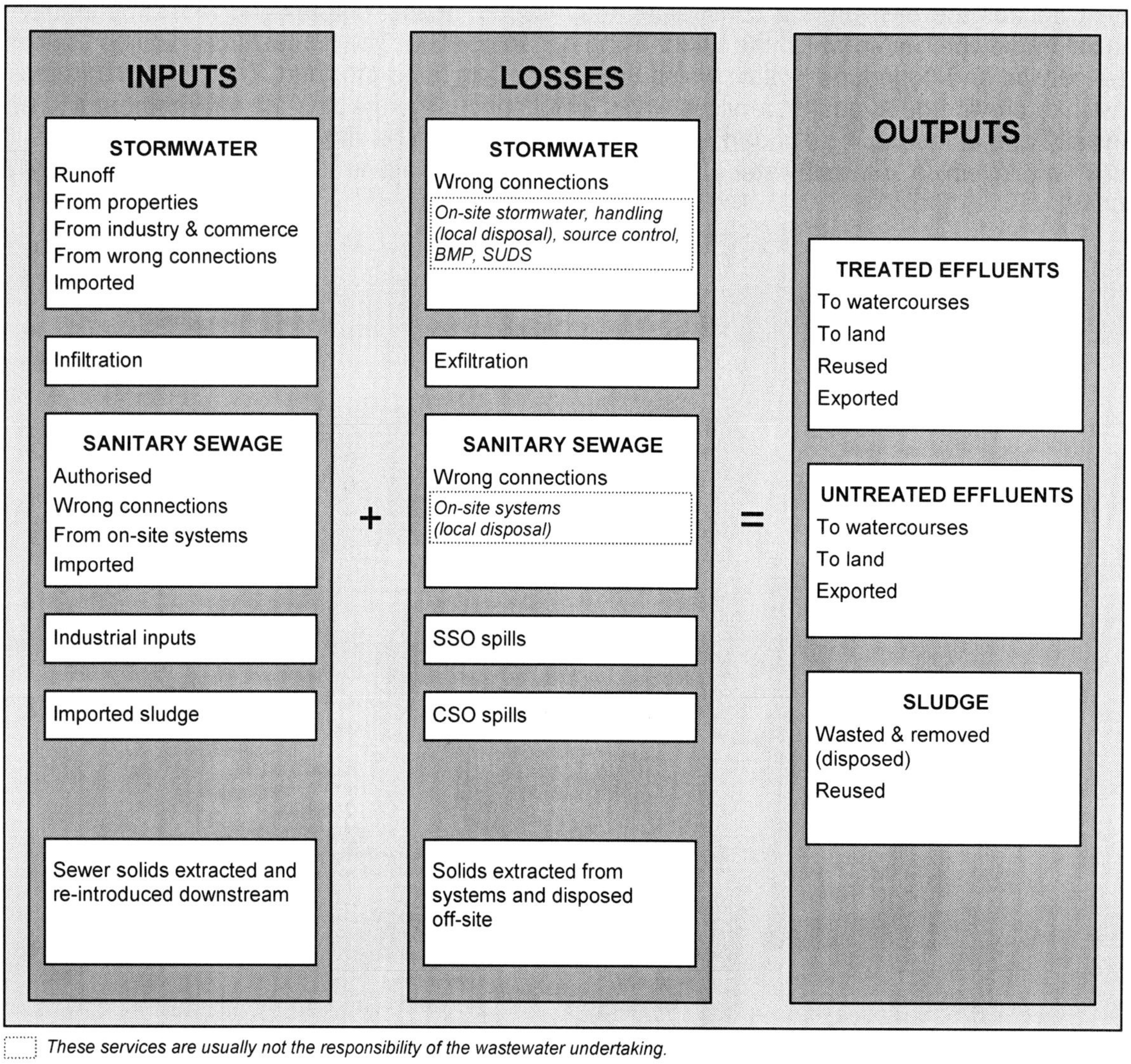

Figure 3. Wastewater balance
(see Appendix 1 for definitions).

The services provided by the undertaking must be related to the users of those services, therefore the inter-relationship between the customers served by the wastewater undertaking and the direct impacts arising from those services are illustrated in Figure 4. The definitions of the terms in this figure are given in the Appendix 1. Figure 4 shows how the various customers (including industry) utilise the services, and how the population and other customer base can be comprised of residents and temporary (visiting) service users. It also shows that parts of the population and other potential customers may manage their own wastewater services on-site or otherwise. The non-users of the services are represented as those not served.

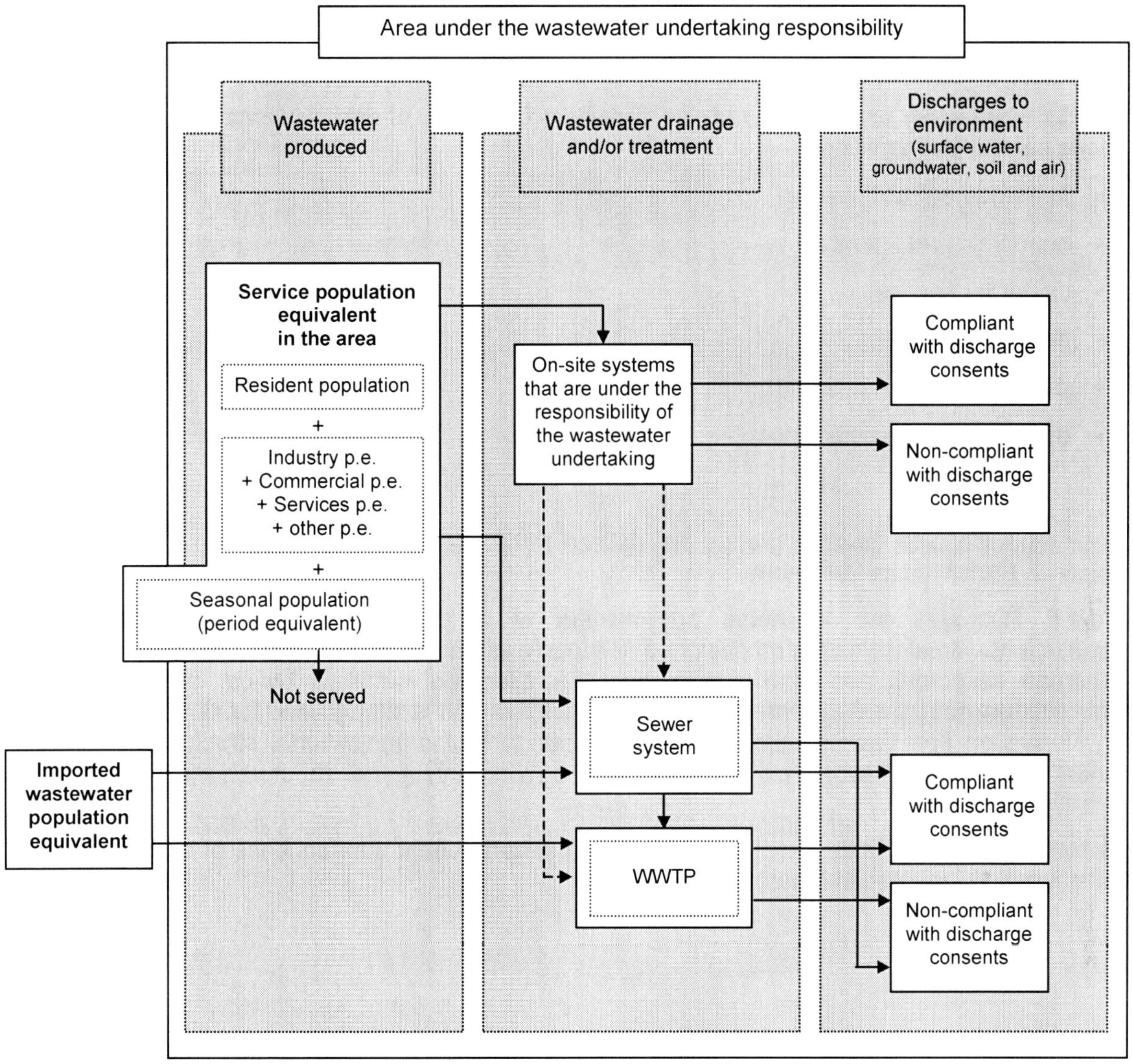

Figure 4. Illustration of service provision in terms of customers and main impacts.

4.5 Organisational functions and definitions

In this Section the key organisational functions of typical wastewater undertakings are illustrated, together with summary descriptions of the various sub-functions of parts of a typical organisation. These are also related to personnel utilisation.

The mains goals of a typical undertaking are:

- to provide a service
- to protect the environment
- to be financially sustainable

A typical wastewater undertaking may be considered in terms of organisational functions under the various component services:

- management and support;
- economic and financial;
- customer service;
- technical services:
- planning and construction;
- operation and maintenance;

These organisational function groups are utilised in the definitions of the PIs in Section 7, which deals with Performance Indicators.

Figure 5 illustrates the functional organisation of a typical wastewater undertaking. This organisational structure has been designed fictitiously as an example to demonstrate the disposition of human resources and how costs might be allocated within a typical organisation. This organisational structure is not intended to be proscriptive and is simply used for illustrative purposes. The PI system has been developed to be independent of organisational structure and therefore manual users do not need to consider how their own organisational structures may differ from this illustrative example.

The following tables (Table 1 to Table 6) specify the component sub-functions of each function and partial function indicated in Figure 5.

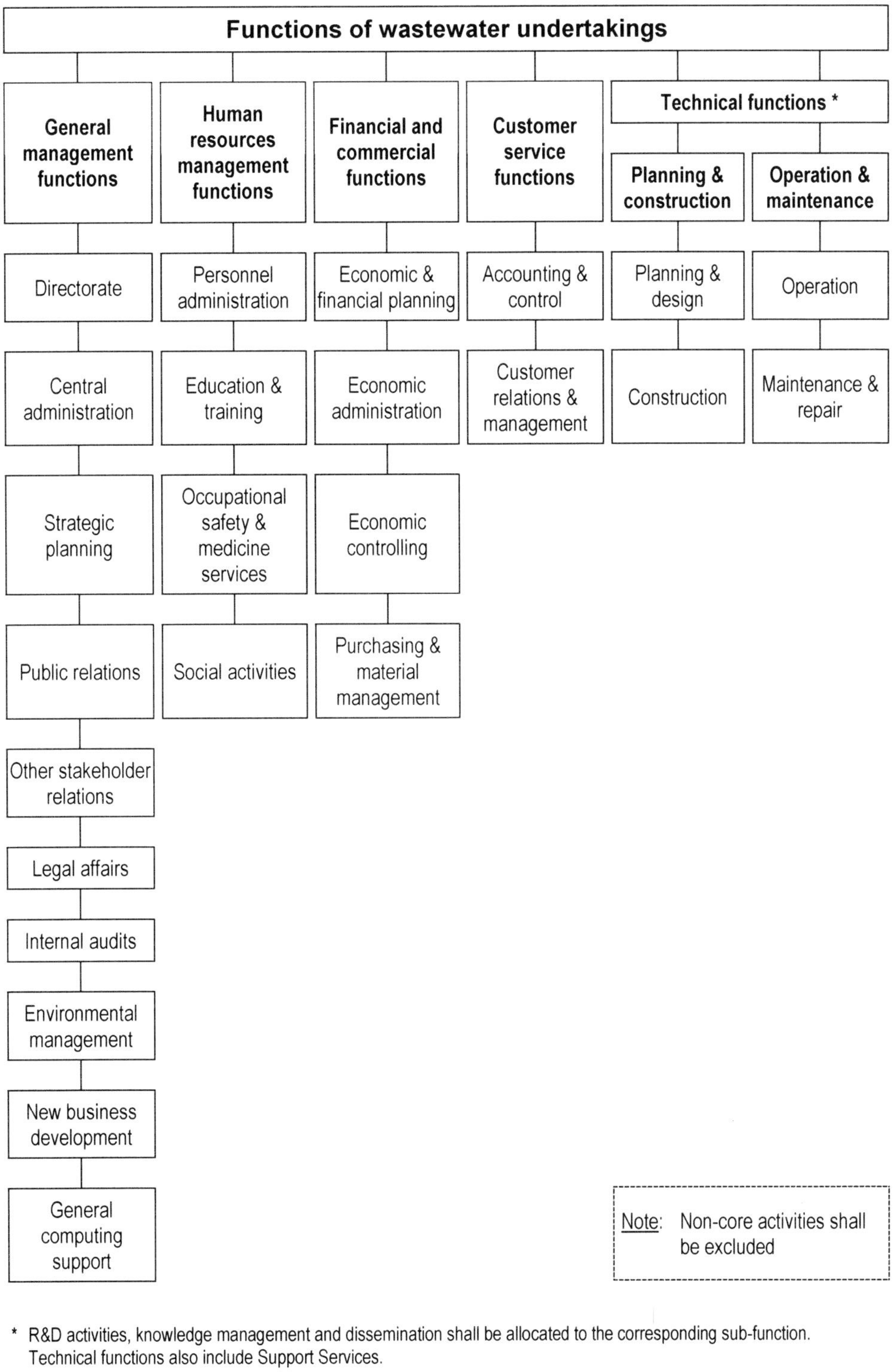

Figure 5. Fictitious example of a wastewater undertaking organisational structure (adapted from Alegre *et al.*, 2000).

Table 1. Organisational functions - General management

MAIN FUNCTIONS	PARTIAL FUNCTIONS	SUB FUNCTIONS
General management functions	Directorate	❑ Strategic policies ❑ External relationships ❑ New business
	Central administration	❑ General secretary ❑ Archiving
	Strategic planning	❑ Objectives definition ❑ Organisation ❑ Information systems planning ❑ Performance assessment
	Public relations [1]	❑ Press service and releases ❑ Communication and image ❑ Customer surveys ❑ Public awareness campaigns ❑ Marketing [2] ❑ Publications and information ❑ Company reports
	Other stakeholders relations	❑ *shareholder* affairs ❑ Policy-making bodies affairs ❑ Regulatory affairs ❑ Other *stakeholders* affairs
	Legal affairs	❑ Legal contracts ❑ Tenders documentation ❑ Legal protection and insurance ❑ Land administration ❑ Contentious matters and litigation
	Internal audits	❑ Check of financial procedures ❑ Check of administration procedures ❑ Check of technical procedures ❑ Quality controls
	Environmental management	❑ Environmental policies ❑ Environmental management and audits ❑ Control of waste management
	General computing support	❑ Management and operation of central computing resources and internal networks ❑ General management of the decentralized computing resources ❑ Software management, including intranet and internet use ❑ Design, implementation and management of information systems, excluding specific applications for technical functions ❑ Training and technical support to the undertaking users
	New business development	❑ Expansion of existing customer base ❑ Expansion of services provided

1 Public relations: interfacing activities with the public. Includes marketing, but unlike marketing is not necessarily intended to increase sales or promote image. Includes public meetings, public participatory activities, customer councils and advisory groups, educational programmes. Also includes stakeholder relations where these relate to the community.

2 Marketing: comprises a wide variety of functions such as advertising, mail-shot, retailing and merchandising, sales, wholesaling, market research, and pricing goods, related to the distribution and sale of goods or services. The aim of marketing is to inform, acquire, retain and satisfy customers.

Table 2. Organisational functions - Human resources management tasks

MAIN FUNCTIONS	PARTIAL FUNCTIONS	SUB FUNCTIONS
Human Resources management	Personnel administration	❑ Personnel planning and development ❑ Recruitment ❑ Personnel management ❑ Job evaluation ❑ Payroll accounting and payment
	Education and training	❑ Planning of training and staff development ❑ Implementation and control of training
	Occupational safety and medicine services	❑ Medical checks ❑ Evaluation of unfitness for work ❑ Working safety policies, implementation and control
	Social activities	❑ Integration of new personnel ❑ Social assistance ❑ Socio-cultural activities
	Management of tangible assets	❑ Vehicle management ❑ Maintenance of buildings and equipment

Table 3. Organisational functions - Financial and commercial tasks

MAIN FUNCTIONS	PARTIAL FUNCTIONS	SUB FUNCTIONS
Financial and commercial tasks	Economic and financial planning	❑ Investment planning and control ❑ Cost planning and control ❑ Revenue and cost recovery planning ❑ Financial planning and policies
	Economic administration	❑ Book keeping ❑ Tangible assets assessment ❑ Annual accounts reporting
	Economic controlling	❑ Planning control ❑ Accounting control ❑ Economic performance assessment
	Purchasing and material management	❑ Suppliers information management ❑ Advertisement ❑ Commissioning ❑ Warehouse management ❑ Vehicle maintenance and services

Table 4. Organisational functions - Customer service

MAIN FUNCTIONS	PARTIAL FUNCTIONS	SUB FUNCTIONS
Customer service	Accounting and control	❑ Accounting ❑ Collection control
	Customer relations and management	❑ Incoming order management ❑ Customer consultation ❑ *complaint* management ❑ Information dissemination

Table 5. Organisational functions - Planning and construction[3]

MAIN FUNCTIONS	PARTIAL FUNCTIONS	SUB FUNCTIONS
Planning and construction	Planning and design	❑ Discharge management and reception bodies protection ❑ *wastewater* collection, transportation and treatment master plan, with medium and long term demand management ❑ Asset management planning and *rehabilitation* programmes ❑ Planning, design and calculation of plants, networks and equipment ❑ Specification of plants, construction and materials selection ❑ Planning and design of technologies e.g. information technologies ❑ Procurement of external design services in relation to the above ❑ Statistics ❑ Research and development ❑ Safety and risk assessment
	Construction	❑ Procurement of construction works by external contract ❑ Direction of works undertaken by direct labour ❑ Supervision, quality control and acceptance ❑ Commissioning of new plants, networks and equipment ❑ Payments and financial control

3 Planning and construction is the set of tasks related to planning and provision of new assets and relining, replacement, or reinforcement of existing assets.

Table 6. Organisational functions - Operation and maintenance

MAIN FUNCTIONS	PARTIAL FUNCTIONS	SUB FUNCTIONS
Operation and maintenance	Operation	❑ Supervision of source protection areas ❑ Planning of operation, monitoring and control ❑ Standby duties and risk management ❑ Documentation, data collection, record keeping and data processing (including management and operation of information systems) ❑ Wastewater quality monitoring, sampling and analysing (laboratories) ❑ *CSO* and influent management ❑ Operational acceptance of new plants, networks and equipment ❑ Re-commissioning of systems after shut down
	Maintenance and repair	❑ System *inspection* ❑ Regular maintenance checks or service activities ❑ *repairs* of failures and other defects ❑ *cleaning* ❑ *refurbishment* ❑ Regular meter *replacement* and maintenance

4.6 Financial aspects

Companies managing wastewater services may operate under a range of types of contractual arrangement and financial regime. Municipalities can retain the ownership of the assets and operate the services directly, including revenue collection. Alternatively, a Municipality, retaining ownership of the assets, can entrust a third party, either a public or a private company (or a partnership of both of these), to carry out any of the services on its' behalf. The Municipality, retaining ownership of the assets, can involve a third party in some or all of the following in accordance with an agreed operating contract: asset maintenance; operation of all or some of the services; collection of revenues. In this case the third party service provider bears only the related operation and maintenance costs, and earns a contracted fee in accordance with the terms and conditions of its contract and service agreement. In a further step towards private sector participation, the third party can have the right to use the assets to provide the services, bearing also the costs of its *rehabilitation* and *renewal* and eventually the construction of new assets, by means of a concession contract with the Municipality. In this case most of the risks in providing the service are transferred to the private contractor.

An alternative is that the Municipality, by means of a license, transfers the assets to a third party, who holds the asset ownership, bearing the costs of maintenance, *renewal* and construction of any new assets, and the operation costs. This system may entail collection of revenues and payment by the contractor to the Municipality.

In a number of countries the wastewater (and often also the water) infrastructure assets are entirely owned by a private company, and in some cases even the water resource system is owned privately. Such companies may choose to operate and manage all aspects of their own service provision, or utilise specialists to sub-contract aspects of their activities, as above. The various Water Service Providers in England & Wales have asset ownership, but each operates using a different selection of contracted out services.

Increasingly worldwide wastewater services are operated and managed by large multi-national entities that have developed particular expertise in such service provision. A common form of this system is that where the resource and asset ownership is retained by the Municipality but service provision is undertaken by a third party. A good example of this is in France where two large companies operate most of the water and wastewater systems on behalf of the French Municipalities.

Many other combinations of public-private system can be adopted. The following tables, definitions and indicators are related to the case in which only one entity, the Municipality or a third party (the undertaking), holds the ownership of the assets and operates the services, bearing the related capital and running costs and obtaining the related revenues from the customers.

It is also presumed that the undertaking has responsibility for the whole system, from the reception of the wastewater, its collection, transportation and full treatment and disposal, including related solids and *sludge treatment* and final disposal. The wastewater system can also be combined *stormwater* and *sewage*, separate (only sewage) or partially combined. What are not considered here are the parts of the system that are not operated by the undertaking. This may include on-site (local owner) sanitation, stormwater *source control*, *BMP* or *SUDS* systems.

The undertaking should utilise an activity base costing system for its accounts, showing the costs divided by each activity, and a Profit and Loss statement referred to each component service. This accounting system is assumed also to distinguish the values of the investments related to each service provided, with an appropriate Balance Sheet. For the purposes of this manual this can be directly accounted or estimated where information is not available. A multi-service company (providing services for water supply, wastewater and wastewater treatment) should therefore be able to calculate each financial PI related specifically to each service activity. Where the undertaking does not have an activity-based costing system, estimation will be needed based on generally accepted accounting criteria, such as FTE: Full Time Equivalent, etc.

It is essential that a consistent accounting approach be utilised to determine each PI where they are to be used for comparative purposes.

The financial definitions and annual profit and loss architecture and the investment definitions are presented in the following diagrams and tables. The annual changes in financial position and the structure of the balance sheet at the end of the accounting period must relate only to the aspects of the undertaking's services used to provide wastewater services. If the financial statements are only available in a form that is consolidated together with other services, this should be clearly stated in order to avoid any spurious comparisons.

Table 7. Annual profit and loss architecture scheme
(adapted from Alegre *et al.*, 2000).

Annual costs (wG5 =wG6+ wG7)

- Running costs (wG6=wG8+wG9-wG33)
- Capital costs (wG7=wG26+wG29)

Service revenues (wG3) Work in progress (-) Capitalised costs of self-constructed assets (wG33) Other operating revenues (-)	Total Operating Revenues (wG2)
Operational costs (wG8): ⇒ External services (wG10) ⇒ Energy (wG11) ⇒ Chemicals, other consumables and materials for maintenance and repair (wG12) ⇒ Leasing and rentals (wG13) ⇒ Taxes, levies and fees (wG14) ⇒ Exceptional earnings and losses (wG15) ⇒ Other operating costs (wG16) Internal manpower costs (wG9)	Operating costs (wG8+wG9)
[SE + WWT] E.B.I.T.D.A. (wG50+wG26) Earnings before interests, taxes, depreciation and amortisation	
Depreciation (referred to the book values) (wG26)	
[SE + WWT] E.B.I.T. = O.I. (wG50 = wG2 -wG8-wG9-wG26) Earnings before interests and taxes = Operating Income of the wastewater service	
Interest expenses (wG27) Interest income (wG28)	Net interest (wG29=wG27-wG28)
[SE + WWT] E.B.T. = G.I. Earnings before taxes = Gross Income of the wastewater service (-)	
Income taxes (-)	
[SE + WWT] NET INCOME of the wastewater service (wG52)	

Detailed definitions related to Table 7 are given below:

Table 8. Definitions of terms related to annual profit and loss architecture

ITEM IN Table 7	INCLUDE
ENERGY COSTS	POWER: all energy costs for wastewater pumping and treatment – electricity and fuel for motive machinery.
EXTERNAL SERVICES	OUTSOURCING: outsourcing of technical or administrative services, such as consultants, contractors undertaking operational tasks, and accounting fees. SOFTWARE LICENCES AND IT SUPPORT: licence fees on computer software and technical support by software companies. ASSOCIATED COMPANIES: costs of associated companies that are not included in other items. SERVICES TO THIRD PARTY: operating costs of providing services to third parties (other than the entrusted and/or wastewater treatment function) that are not included in other items: i.e. *sludge treatment*, *septic tanks cleaning*, etc.
LEASING AND RENTALS	Payments for leasing or renting premises, vehicles, mobile and fixed plant and equipment.
PURCHASES	CHEMICALS, OTHER CONSUMABLES AND MATERIALS: all materials, chemicals and other consumables apart from energy, that are not in HIRED AND CONTRACTED SERVICES and which are required for operation of *sewer systems* and treatment plants
TAXES, LEVIES AND FEES	Any operating licence paid to a governmental or municipal authority others than income taxes.
EXCEPTIONAL EARNINGS AND LOSSES	Any exceptional income or expenditure from donations, investment subsidies, compensations or adjustments related to sales / writing-off of fixed assets.
OTHER OPERATING COSTS	OTHER DIRECT COSTS: any other operating costs (but excluding interest and taxation, in aggregation). GENERAL AND SUPPORT EXPENDITURES: the aggregate direct cost of GENERAL AND SUPPORT ACTIVITIES (manpower costs excluded) (see section 4.5 for definitions). CUSTOMER SERVICES: costs directly associated with customer services that are not included in previous items, related to customer accounting, reading of wastewater flowmeters, debt recovery, customers' enquiries and *complaints* handling. SCIENTIFIC SERVICES: costs directly associated with scientific and laboratory services and with monitoring activities that are not included in previous items. OTHER BUSINESS ACTIVITIES: costs directly associated with other business activities that are not included in previous items, except for cost depreciation. DOUBTFUL DEBTS: charge/credit to the profit and loss account for bad and doubtful debts.
INTERNAL MANPOWER COSTS	EMPLOYMENT COSTS: the sum of the total manpower costs of permanent and temporary personnel, including employment-related social costs and benefits paid by the employer.

ITEM IN Table 7	INCLUDE
CAPITALISED COSTS OF SELF-CONSTRUCTED ASSETS (negative allocation)	The summation of the amounts in each of the above cost categories that have been incurred in the construction of new or *rehabilitated* assets, including work in progress. *For running cost PI calculation purposes, capitalised costs of self-constructed assets are to be apportioned negatively against each of the above running cost items having due regard to the strict observance of the following relationship:* Running costs from the Annual Profit and Loss Architecture = Operation and Maintenance Costs + Internal Manpower Costs – Capitalised costs of self-constructed assets = Net value of: (Energy + External services + Purchased merchandises + lease rentals + operating expenditures + internal manpower costs)
DEPRECIATION COSTS (ON BOOK VALUES)	COST DEPRECIATION: cost depreciation charge on tangible fixed assets AMORTISATION OF INTANGIBLE ASSETS: any amortisation or other reduction in the balance sheet valuation of intangible assets, such as goodwill. THIRD PARTY SERVICES: cost depreciation on assets relating to third party services, together with any infrastructure *renewal* charge for infrastructure assets relating to third party services.
NET INTEREST COSTS	NET INTEREST: Net cost of short, medium and long-term loan capital (INTEREST EXPENSES – INTEREST INCOME).
INCOME TAXES	TAXES: All taxes and levies on gross income related to *sewer system* and wastewater treatment activities. *Tax costs and levies strictly connected with wastewater works operations (such as sewage and concession charges, environmental levies, water control authority charges, etc.) have to be regarded as operational costs and included in TAXES, LEVIES AND FEES (RUNNING COSTS).*

Table 9. Investment definitions

TERM	DEFINITION
INVESTMENTS SUBJECT TO DEPRECIATION	Investments that have to be subject to depreciation as listed in detail by the general accepted accounting principles (IASC)
DEPRECIATED HISTORICAL VALUE OF TANGIBLE ASSETS	Sum, over all the tangible assets, of the depreciation value applied since the date of purchase.
HISTORICAL VALUE OF TANGIBLE ASSETS	Refers to the gross book value of total expenditures (that appears for the net value in Table 11) in infrastructure and non-infrastructure tangible assets.

Table 10. Annual changes in company financial position.

APPLICATION OF FUNDS	SOURCES OF FUNDS
Capital expenditures (Investments) [4] **:**	Internal sources from ***cash-flow:***
⇒ intangible assets	⇒ retained earnings (after dividends and other payments out of net income)
⇒ licences and rights	⇒ depreciation
⇒ tangible assets	External sources:
⇒ *sewer system*	⇒ capital receipts
⇒ plants and machinery	⇒ borrowing and loans obtained
⇒ other equipment	⇒ sale of fixed assets
⇒ financial assets	
⇒ shares in other companies	
⇒ other financial investments	
Financing:	
⇒ borrowings and loans repaid	

⇩

Company Working Capital increase [5]	Company Working Capital decrease [5]

Table 11. Structure of the company balance sheet at the end of the year.

ASSETS	SHAREHOLDERS' EQUITY AND LIABILITIES
Intangible assets	*shareholders'* Equity
⇒ *goodwill*	⇒ subscribed share capital
⇒ net value of licences and rights	⇒ capital reserves
	⇒ other reserves
	⇒ net income for the year
Tangible assets	Long term liabilities
⇒ net value of *wastewater treatment plants*	⇒ bonds
⇒ net value of other assets	⇒ long term financial debts
Financial assets	
⇒ net value of financial investments	
Current assets [6]	Current liabilities [6]
⇒ cash at bank and in hand	
⇒ accounts receivable from wastewater services	⇒ accounts payable
⇒ other accounts receivable	⇒ current portion of long term debt
⇒ inventories	⇒ miscellaneous current liabilities
⇒ prepaid expenses	

4 Capital expenditures (investment in tangible, intangible and financial assets) is the same as "total investment".

5 The increase or the decrease is the balance between application and sources.

6 Current assets - Current liabilities = Working capital

4.7 *Structure of the PI system*

The previous Sections provide an outline of the systems and organisation for which this manual and PIs are required. The indicators and structure in which they have been defined have been framed in terms of the principal management objectives common to all undertakings, in such a way that the PI system is independent of the state of development or internal organisational structure of any particular service provider. The PI system is structured into six separate categories of performance as illustrated in Figure 6. The interpretation of the performance of an undertaking cannot be carried out without taking into account the context in which it operates. In addition, it is necessary to consider also the characteristics of the infrastructure and resource systems and the characteristics of the region in which the services are provided. Hence the structure of the PI system also includes profiles for context, system and region, as illustrated in Figure 6.

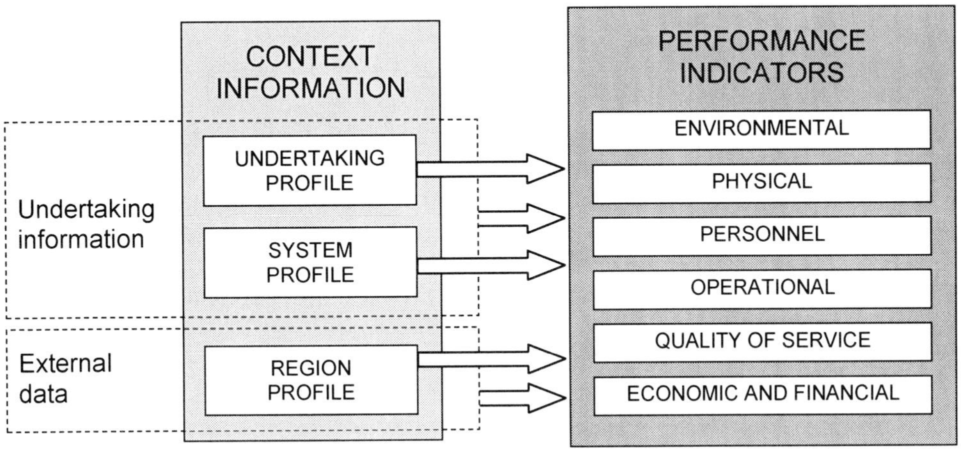

Figure 6. Structure of wastewater CI and PI
(adapted from Matos *et al.*, 2002a).

The undertaking profile outlines the business context in which the organisation operates. The system profile focuses mainly on the wastewater volumes managed, the physical assets, the technological means used to provide the services and on the demographic aspects of the customers. This profile requires more detail than the other system or region profiles because it also contains the descriptive indicators that are essential for the interpretation of the performance indicators. The region profile will be particularly relevant for comparisons between undertakings, as it provides the information needed to understand the demographic, economic, geographical and environmental context within which each undertaker operates. The context, system and region profile information are considered further in Section 8 - Context information. Table 12 shows how the six performance categories are coded for PI labelling. This is further developed in Section 7 - Performance Indicators.

Table 12. Structure of the PI framework

	Code	
Performance indicators	*wEn*	Environmental indicators
	wPe	Personnel indicators
	wPh	Physical indicators
	wOp	Operational indicators
	wQS	Quality of service indicators
	wFi	Economic and financial indicators

The way in which the data collected for the PIs and associated context information presented in this manual may be used to evaluate efficiency and effectiveness of the provision of services by an undertaking is beyond the scope of the manual. This is because performance evaluation is part of *process* benchmarking (Section 1.4) and hence is dependent on the objective of the performance assessment, which cannot here be generalised. The PI system presented in this manual is intended for use for metric benchmarking purposes, but will provide the baseline information required to support process benchmarking.

5. IMPLEMENTATION STRATEGY OF A PI SYSTEM

5.1 Introduction

There are three main phases in the implementation of a PI system as part of a metric benchmarking process. The first phase is to clearly define the objectives via definition of *the strategic performance assessment policy* for the undertaking. This phase requires particular participation and engagement of the senior staff of the undertaking. From this, the objectives of the assessment will be determined and these should be stated clearly in terms that everyone within the undertaking and each *stakeholder* could understand. In addition, a work plan should be formulated to specify who within the undertaking will lead and carry out the assessment and how it will be implemented. It is recommended that a specialist PI team be set up, and a decision made whether or not to use the IWA PI manual. The second phase is the determination of the PIs to be assessed, with PI review, selection, trial data collection, and initial evaluations to determine which of the pre-selected PIs are deemed to be important. The third phase is the detailed implementation of the internal procedures to be used to carry out the PI assessment based on the PIs selected as being those of most importance. The same approach should be used in parallel with the PI processes for the CI data. These three phases, together with their components are shown in Figure 7.

Although the pre-selected PI team should undertake the bulk of the work in producing the PIs and CIs, it is essential that the senior staff of the undertaking be engaged fully in the processes. Reported experience by the six-Cities group, amongst others, shows that PI assessments frequently fail where senior staff are perceived by the majority of the staff in the undertaking and those carrying out the assessment to lack commitment to the process. Failure can arise due to both poor PI team working and reluctance on the part of other staff to provide the requisite data.

5.2 Definition of the strategic performance assessment policy

The implementation of a PI system has to be objective-oriented. The definition of the PI system to be adopted should therefore reflect directly the specific objectives intended in use. Potential objectives and motives for using a PI system are outlined in Section 1.3, and illustrated in Figure 7. The most common application is use as a component of metric benchmarking (Section 1.4). It is necessary to clearly define the potential use of the PIs, as different PI sub-groups will have different significance depending upon the objective of the assessment. For internal review purposes the undertaking may choose not to formally collect CI data, as knowledge about the context within which the organisation operates is held intrinsically within the asset base. CIs are essential only where comparisons are to be made with other organisations, or possibly between sub-groups, possibly regionally diverse, within the organisation.

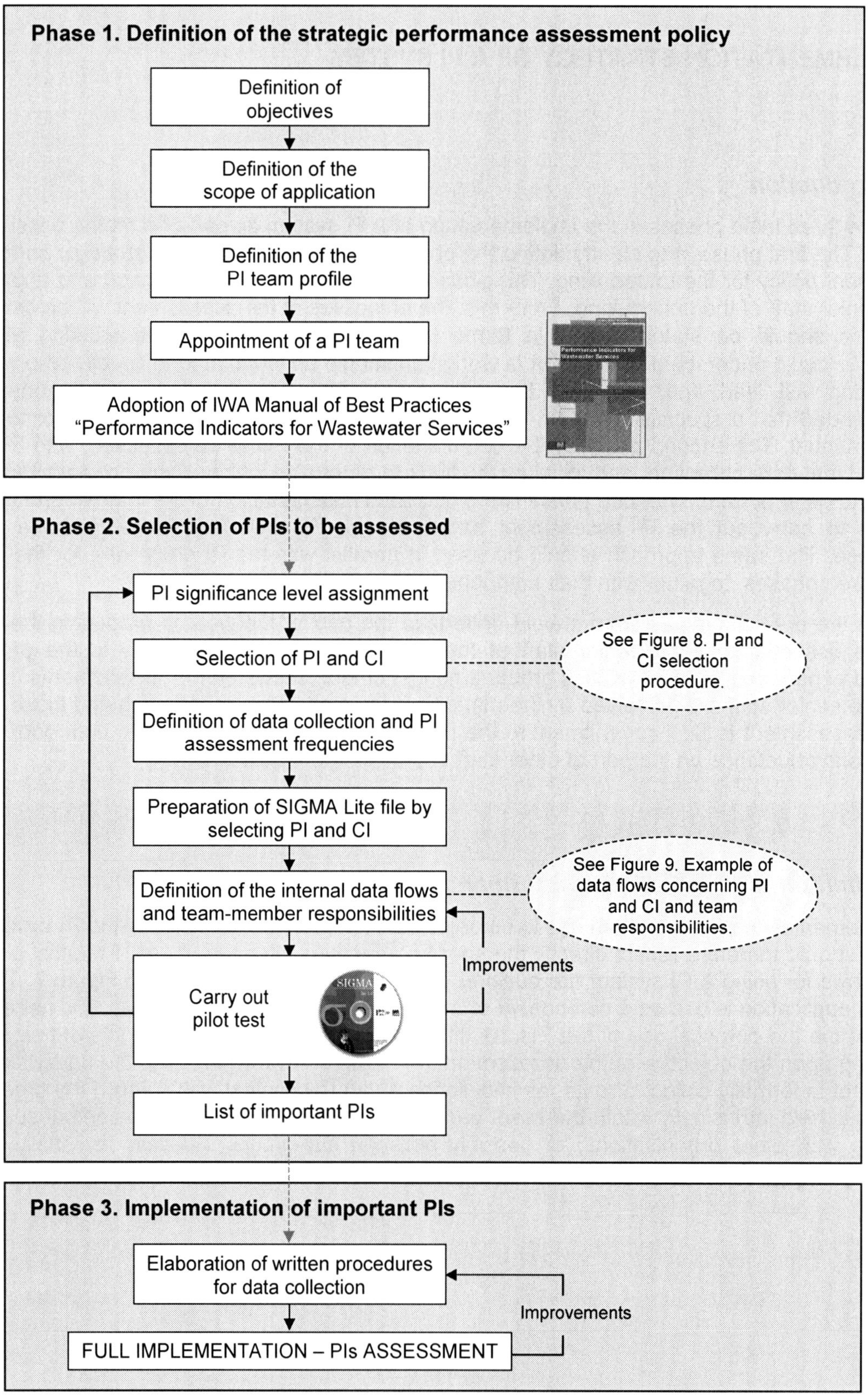

Figure 7. Phases of the PI system implementation process (adapted from Alegre, 2002a).

In Phase 1, the undertaking should ensure that all internal and external *stakeholders* required to properly define the objectives and scope of the PI application are either involved in the definition or are otherwise appropriately included. This will ensure that all relevant stakeholders (those commissioning and driving the process, those carrying it out, those providing the key data and those interested in the outcomes) are effectively engaged. From this, the characteristics and constituents of the PI/CI team, and leader can be formulated. In addition, the undertaking can at this stage, identify which parts of the organisation will have to be engaged in the second, implementation phase. The establishment of the PI team should be based on the principles and objectives defined from the first part of Phase 1. Successful implementation of a PI system within an organisation normally requires that the PI team leader is a senior staff member, if necessary assisted by less experienced colleagues. As stated above, the involvement and active participation of the senior staff, usually the decision-makers who are going to use the PI information, is also essential. It is also strongly recommended that representatives of the departments/sections who need to provide the PI data are core members of the PI team, in order to ensure their commitment to all the stages of the process. Experience from pilot testing the water supply PI manual has demonstrated that an important benefit of the implementation of a PI system may be derived from staff participation in the experience. Participants experience a wider range of alternative views and perspectives from other staff members with different responsibilities and backgrounds. It is essential at this stage to properly brief the PI team about the purpose, objectives and expected outcomes from the establishment of the PI system within the undertaking.

The next stage in the process (Figure 7) is to review available PI systems and select that most appropriate to the specified objectives. Although this manual may be used to generally assist users in the application of a wide range of PI systems, it is assumed that users will ordinarily adopt the IWA manual of PIs for wastewater services. Subsequent guidance in this manual may not be applicable to other PI systems.

5.3 Selection of PIs to be assessed

Phase 2 of implementation deals with the selection of PIs/CIs. The 182 PIs presented in this manual are intended to provide a global set that should cover all types of undertaking and objectives for using a PI system. It is not the intention that undertakings should necessarily use all of the 182 PIs, and it is unlikely that each of the PIs would be important for any one undertaking. Hence, users should carry out a PI selection process. Guidance is provided here to assist users to select the PIs that are particularly relevant for their performance assessment. There are five stages in PI/CI selection:

- Pre-selection of the PIs considered to be of importance for the assessment
- Assignment of significance levels to pre-selected individual PIs
- Selection of the actual PIs and CIs to be used in the initial assessment
- Pilot testing of the initially selected PIs
- Final selection of PIs deemed to be important and to be used in the assessment

The processes to be used for this are illustrated in Figure 8 and Figure 9.

The first step is to assign relative significance levels to each PI depending upon the objective and context of the PI application by the wastewater undertaking. This should be done in consultation with the various stakeholders. It is recommended that the PIs defined in this manual be initially classified by the user into one of three significance levels: *high, medium* and *low.* The undertaking should then initially assess whether or not the PIs of highest significance are to be used in the full implementation, followed by consideration of the PIs of medium and then lower significance. Even where it is known that data are not available or of doubtful veracity, individual PIs should still be ranked according to their overall significance. Hence a further outcome from the PI significance level assignment is to help prioritise where data are needed and hence, where new data collection programmes may be required.

An illustration of the assignment of relative levels of significance to the PIs is given in Appendix 3, based on the perspective of a typical wastewater undertaking in a developed country. This may be considered as an example only, and is not proscriptive for all users or applications. Relative significance levels should be assigned based on key *stakeholder* consultation. Figure 9 shows that initial selection of PIs should be made at senior level within the wastewater undertaking. It would be expected that this would entail consultation with the PI team, who would interact with the suppliers of the data (the various Departments or sub-groups within the undertaking). Additional consultation outside the undertaking may be required where, for example, the performance assessment is being conducted for regulatory or other purposes. An illustration of the participants that may be required in assigning levels of significance and selecting relevant PIs is given in Table 13. The undertaking may not necessarily consult external stakeholders where the evaluation will be for internal usage only.

Table 13. Illustration of types of participants and data implications for the assignment of relative significance levels to the set of PIs given in this manual

Example of the purpose (context) for using PI system	Stakeholders involved in assignment of relative significance to each PI	Data need implications
For internal review purposes only by wastewater undertaking	Senior Management and all levels within the undertaking that are involved in managing, operating the system to be reviewed. As many or as few of the PIs may be assigned to the highest (priority) level as deemed appropriate for internal review.	The undertaking may or may not already have systems in place to collect, process and record the requisite data.
For use as required by an environmental regulator	As above, but also including the regulator and possibly proactive users, such as environmental interest groups (NGOs). It is probable that there will be some negotiation as to the PIs to be considered at the highest level of significance.	It is probable that regulators will require only those PIs to be assessed that are relevant to their individual area of interest, e.g. economic or environmental. It is likely that the undertaker will already be recording the required data.
For interaction with a wide range of customer groups	As wide a range of stakeholders as possible. The undertaking will have first to convince the other stakeholders of the need to prioritise the PIs and report only on those most significant, but it may be necessary to eventually increase the number of PIs reported.	It is unlikely that the undertaking will have all of the data needed for all the PIs perceived to be significant by diverse groups of stakeholders. A more limited set of indicators may already be in place as part of a customer 'council' or other form of representation.

Following assignment of significance levels to each of the PIs, the next stage is to pre-select the PIs to be used in a pilot assessment. Certainly those with the highest significance should be selected. The final selection of the PIs to be used in the full implementation is then an iterative process. The experience of the senior staff should provide the initial selection giving a preliminary listing based on the significance level. Then the definition of the data required and the associated reliability and accuracy may lead to the confirmation or the rejection of each PI to be used in the final assessment. At this stage the required frequency of evaluation of each PI should also be considered (Section 6.4). In general, PIs for the overall assessment of the undertaking should be assessed once per year, but there may be indicators and assessment objectives for which different frequencies are advisable.

The procedure for checking each PI for data availability is illustrated in Figure 8. The CIs should also be included unless the PI exercise is only being conducted for internal review purposes. At this stage the selection will be a combination of both relative significance and data availability. Where a PI is of high significance, but data are not available or acquisition is difficult, the undertaking should still include the PI in the selection, but put in place measures to obtain the data for future PI assessments.

For some assessment needs it may be necessary for the Undertaker to customise the core PIs. These may need to be supplemented with expanded or more detailed sub-indicators. Selection processes for these should follow the same procedure as here. Following PI pre-selection, it is recommended that an individual, department or sub-group be assigned to be responsible for data provision, validation and confirmation for each PI.

By working successively through the pre-selected PIs in order of perceived significance level, each one can be considered for inclusion or rejection for the final list of the most important PIs. The PIs included will then be those used in the full assessment, implemented in Phase 3. It may be that via the pilot testing, a PI deemed to be significant in the pre-selection stage is found to be not as important as initially envisaged, and may therefore not be included in the final selection. It is also possible that PIs initially ranked as of medium or low significance, may in fact be found to be important, although not necessarily particularly significant. Hence these may need to be included within the finally selected list for detailed assessment.

All of the PI team members should participate in the process of interpretation and final definition of the PIs, including reporting frequency for each. Ultimately it will be these staff that assist in the identification and implementation of procedures where comparisons and evaluation of the PIs indicate a lack of service effectiveness or efficiency. This phase is critical to ensure that staff is motivated.

The PI implementation team has to identify all the data required and work with the data providers in order to assess data availability, reliability and accuracy. Whenever data for a PI is not available or is inaccurate, senior staff, the PI team and data providers have to decide whether or not to improve the data by further work or to accept that the current indicators cannot be assessed properly or are of doubtful accuracy. Wherever possible synergy with the existing information systems that the undertaking operates should be sought. Data providers need to be confident that the PI system is part of the normal activity of the organisation, and that there are major advantages in using the same data (i.e. defined and expressed in the same way) for the various applications requiring similar information. However, the effective integration of existing data banks is part of a more general strategic objective of most undertakings, and not a PI-specific problem.

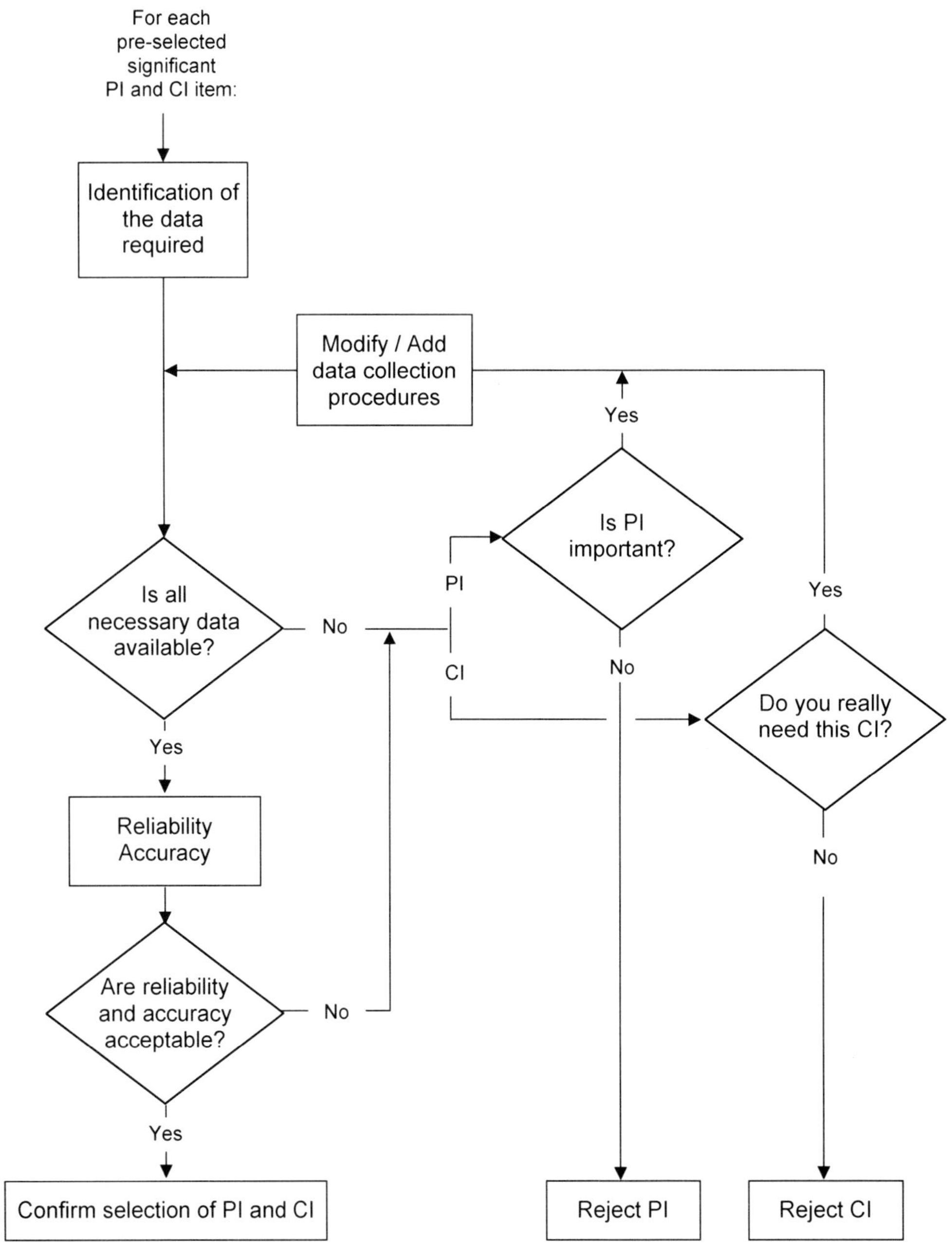

Figure 8. PI and CI selection procedure.

The assessment of data availability, reliability and accuracy, together with the relative significance level for each PI allows for testing, evaluation and consolidation of the PIs finally selected for assessment in the iterative process (Figure 7). This stage will also indicate where there may be opportunities for improvements in the data measurement and collection procedures.

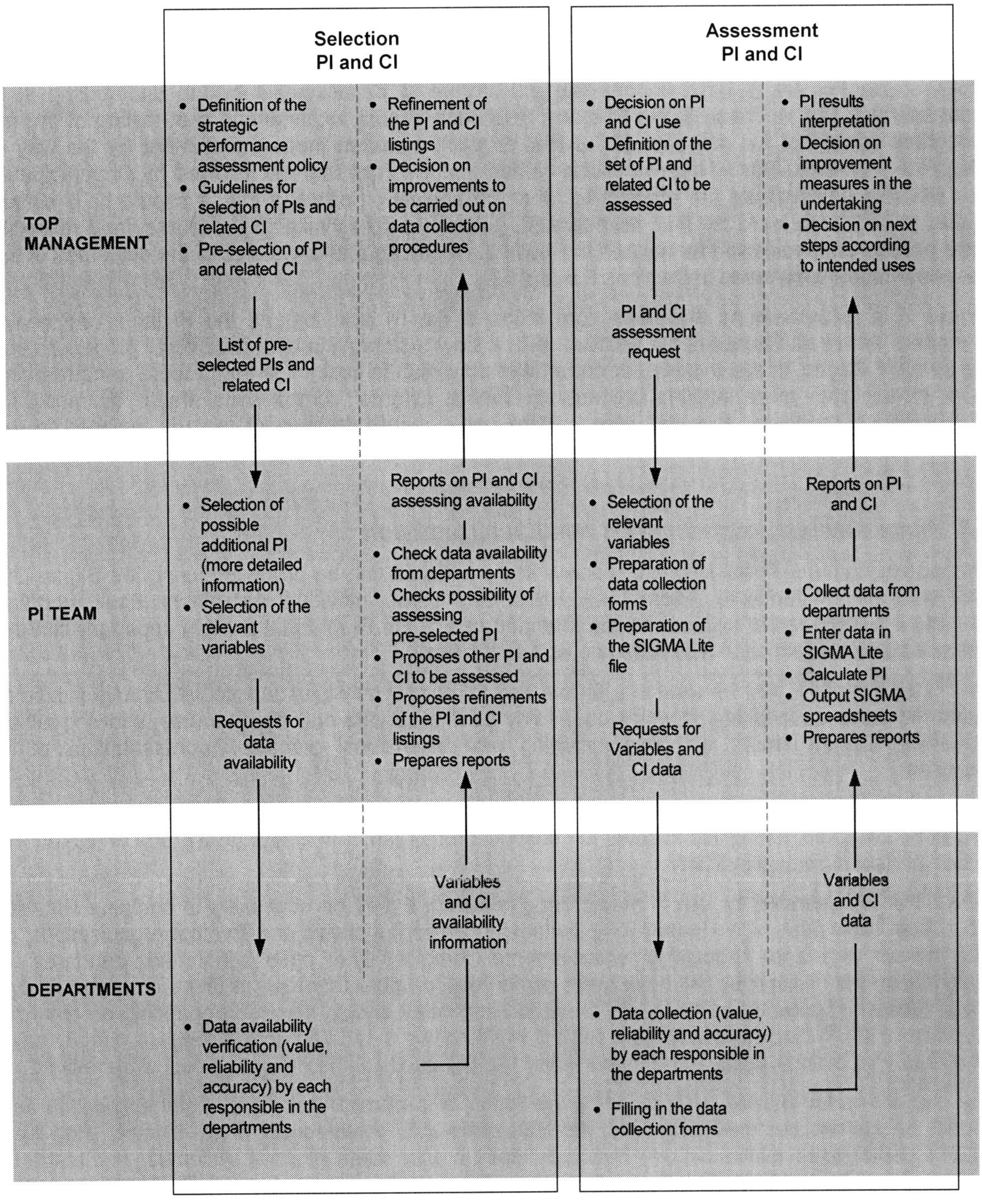

Figure 9. Example of data flows concerning PI and CI and team responsibilities.

Figure 9 shows how the various participants in the PI implementation process interact, together with data flows.

5.4 Implementation of important PIs

This phase deals with the final implementation of the PI assessment process. Phase 2 has defined which of the PIs are of most importance and require to be assessed by the undertaking. It is recommended that in Phase 3, the PI team, who have already experienced pilot testing of the full procedure as part of the selection of the PIs, produce a written method statement for the way in which the final evaluation of the PIs will be carried out. This can then be followed by all participants and will also demonstrate the process to other *stakeholders*. Following this, it should be a simple matter to fully implement the final assessment, given that data availability, accuracy and reliability have already been reviewed for each PI in Phase 2. Reporting and evaluation of the outcomes of the assessment are considered in Sections 5.5 and 5.6.

Finally, it is recommended that upon completion of the PI assessment, the PI team and senior managers review all Phases of the process, with a view to improving both the overall procedure and the detailed stages. In many cases an undertaker may wish to customise the process to correspond more closely with other auditing procedures. Where external comparisons are to be made for benchmarking purposes, this may also require some reconsideration of the PIs included in the assessment.

5.5 PI assessment, reporting and result interpretation

The results from the PI assessment process in this manual can be produced using the Sigma-Lite WW software (Appendix 4). Alternatively users may utilise individual databases. Final reporting should be customized to account for the intended use of the PI information. It is important that the relevant context information is included as well as the PIs.

The PI assessments may be used in a simple way to monitor quality of service with changes in time, within the same undertaking, in which case fairly general reports may be adequate, showing tables containing the PI results, and also including two dimensional graphical representations of PI evolution.

The process of result interpretation usually requires comparison with a reference value or a reference threshold, taking into account the context information, and analysing a group of related PIs rather than an individual indicator.

Where PIs are intended for use in metric benchmarking, it may be necessary to compare PIs with those from other parts of the undertaking performing similar functions, or with other organisations, or with pre-set targets for financial or environmental compliance, or other benchmark standards. In many cases the PI exercise will have been carried out for more than one of these purposes. There are a number of possibilities as to which metric benchmark should be used depending on both the objective of the PI assessment and the context in which this is set. The undertaking will already have identified what is appropriate at the outset when the objectives of the PI assessment were defined.

The use of PIs for internal audit by the undertaking is in general the most useful application and should be carried out over long time periods, preferably continuously once initiated. However, internal performance evaluation only becomes feasible after some years of systematic use of the PI system. There has to be sufficient data for temporal comparisons to be meaningful. Therefore, in the early stages of implementation it may be more feasible to undertake comparisons with the performance of other undertakings or other external reference values. There are often considerable additional benefits that may accrue from inter-comparison with other similar organisations.

Where cross-organisational benchmarking is carried out, individual undertakings may wish to safeguard the confidentiality of some of their PIs. Under these circumstances comparisons of PIs can be made anonymously within carefully selected sub-groups, or managed by an independent external entity that collects the PI data on behalf of the various undertakings, and then processes and reports the information in a form that ensures that individual undertaker identification is not possible.

5.6 Definition and implementation of improvement measures

The definition and implementation of service improvement measures, although not a part of the actual performance assessment, is fundamental to the PI assessment process, as it is usually the reason why PI assessment has been carried out. Weaknesses in service delivery may be identified from the PI process, and a diagnosis should establish the causes and indicate possible remedial measures. Once remedial measures have been identified, undertaking managers can establish short, medium or long-term targets for improvement and implement the remedial measures. The continuing systematic use of the PI system will provide the means to monitor the effectiveness of the measures for benchmark comparison. Any deviations can be investigated in order to introduce appropriate corrective measures.

6. DATA REPORTING

6.1 Confidence grading scheme

A *confidence-grading scheme* is required so that users of the Performance indicators and context information are aware of the reliability of the information available. A confidence grading scheme has been developed for use in England and Wales by the Office of Water Services (OFWAT, 2002), and is now being applied successfully in a number of countries worldwide. This scheme is also recommended for use in this manual as outlined in the subsequent sections.

6.2 Confidence grades

The confidence grades provide a rational method for undertakings to qualify the reliability and accuracy of information provided for the PIs.

It is essential that undertakings and their auditors give proper care and attention to the assignment of appropriate confidence grades to PI data.

A quality-assurance approach that is used to assign confidence grades to each PI is recommended in the methodology. This is particularly important where sampling techniques are used and data are not continuous. Confidence grades should be allocated to reflect the actual and current status of the data, not the future status that it may be intended to achieve (Turrell, 1995). With a continuous PI system, the confidence grades may be used to show how the quality of the base data has developed with time.

Reliability bands:

A - HIGHLY RELIABLE

- Actual — Data based on sound records, procedures, investigations or analyses that are properly documented and recognised as the best available assessment methods.
- Forecasts — Based on extrapolations of high-quality records covering or applicable to 100% of the undertaking's area, kept and updated for a minimum of five years (the forecast will have been reviewed during the reporting period).

B - RELIABLE

- Actual — Generally as in band A, but with minor shortcomings, e.g.: some of the documentation is missing, the assessment is old, or some reliance on unconfirmed reports or some extrapolations are made.
- Forecasts — Based on extrapolations of records covering or applicable to more than 50% of the undertaking's area, kept and updated for a minimum of five years. The forecast will have been reviewed during the previous two years.

C - UNRELIABLE

- Actual — Data based on extrapolation from a limited sample for which band A or B is available.
- Forecasts — Based on extrapolations of records covering more than 30% of the undertaking's area. The forecast will have been reviewed during the previous five years.

D - HIGHLY UNRELIABLE

- Actual — Data based on unconfirmed verbal reports and/or cursory inspections or analysis.
- Forecasts — Based on extrapolated information not complying with bands A, B or C.

Accuracy bands:

Accuracy is defined as the approximation between the result of a given measurement and the (conventionally) correct value for the variable to be measured. The accuracy bands presented below are based on the system adopted in England and Wales. They are to be applied to the measurement and not to the measuring equipment - for example, in some cases the equipment may be highly accurate but is used out of range. Whenever the measurement accuracy cannot be assessed, it should be graded as greater than 100%.

The accuracy bands suggested are:

1. Better than or equal to +/- 1%
2. Not band 1, but better than or equal to +/- 5%
3. Not bands 1 or 2, but better than or equal to +/- 10%
4. Not bands 1, 2 or 3, but better than or equal to +/- 25%
5. Not bands 1, 2, 3 or 4 but better than or equal to +/- 50%
6. Not bands 1, 2, 3, 4 or 5 but better than or equal to +/- 100%

X. Values which fall outside the valid range, such as > 100%, or small numbers.

Overall confidence grades:

The confidence grades will be an alphanumeric code, which couples the reliability band and the accuracy band – for instance:

A2 - Data based on sound records etc. (Highly Reliable, Band A) which is estimated to be within +/- 5% (Accuracy band 2).

C4 - Data based on extrapolation form a limited sample (Unreliable, Band C), which is estimated to be within +/- 25% (Accuracy band 4).

The reliability and accuracy bands would form the matrix of confidence grades shown below:

Table 14. Matrix of confidence grades

Accuracy Bands (%)	Reliability bands			
	A	B	C	D
[0; 1]	A1	++	++	++
]1; 5]	A2	B2	C2	++
]5; 10]	A3	B3	C3	D3
]10; 25]	A4	B4	C4	D4
]25; 50]	++	++	C5	D5
]50; 100]	++	++	++	D6

Note: '++' indicates confidence grades that are considered to be incompatible

Confidence grades should be assessed for every water undertaking and for every indicator, and the corresponding results reported as follows:

Table 15. Reporting of confidence grades (c.g.) for a sequence of years

Year	Result	*'short description'* PI1	*'short description'* PI2	*'short description'* PI3	*'short description'* *PI4*	*'short description'* *PI5*	*'short description'* *PI...*
1996	*PI*	*20%*	*87%/year*	*34/m³*	*40/km sewer*	*55%/year*	*45%*
	c.g.	*B3*	*[C4]*	*A2*	*A1*	*B3*	*A3*
1997	*PI*	*21%*	*86%/year*	*30/m³*	*45/km sewer*	*65%/year*	*45%*
	c.g.	*A2*	*B3*	*A1*	*B1*	*B2*	*B2*
1998	*PI*	*23%*	*84%/year*	*31/m³*	*42/km sewer*	*67%/year*	*45%*
	c.g.	*A1*	*C2*	*A1*	*A3*	*[B4]*	*A2*
1999	*PI*	*23%*	*85%/year*	*26/m³*	*40/km sewer*	*68%/year*	*45%*
	c.g.	*B2*	*A3*	*A3*	*C1*	*C2*	*C1*
...	...	...	...	...	...	...	...

Note: '[ ]' means that the original values were corrected by the audit

To make it possible for comparisons to be carried out between undertakings, confidence grades must be chosen appropriately and applied consistently. The descriptions outlined above are minimum values expected, to achieve the quality of information stated.

Clearly, "A1" or "A2" confidence grades can be achieved for some input variables, although they may not be generally attainable. Undertakings are encouraged to aim for a grade of "B2" (i.e. good or excellent) or better.

6.3 *Confidence grades examples*

Only few data are eligible to be classified "A1", typically only information duly audited and likely to be automatically counted by a computerized system (e.g. number of *complaints*, number of bills issued). Certain economic PIs taken directly from the audited accounts may also be rated "A1". Examples may be the annual cost of investment or the service revenues. However, difficulties in understanding the confidence grading system, or a too optimistic attitude while evaluating the reliability and accuracy of data, may result in a tendency to assign inappropriate "A1" confidence grades.

Other data may be based on high quality and audited records, but the accuracy of the measurement may not be precise. Examples include the determination of the different rates of effluent flow or the volume of *sludge treated*. In these cases, "A2" to "A4" ratings would be appropriate.

For the PIs dealing with the physical buried assets, where assessments of their length, type of material or age are required, there may be historic (and usually) unverified records, unconfirmed reports or extrapolations that require "B", "C" or even lower reliability grading. Time equivalent personnel data per function are also likely to be assigned these grades.

Some undertakings may not yet have any records for particular PIs, but may be able to extrapolate estimates from other assessed information, until such times as a more precise determination can be made. In this case "D3" to "DX" ratings will be applicable.

As many PIs are determined from the combination of two or more individual data items, these PIs cannot be assigned a higher confidence grading than that which would be assigned to any of the individual data items. The PI grading should thus be no better than the lowest of those assigned to the constituent individual data item.

6.4 Periods for data assessment

In Section 5.3, the way in which the PI system should be implemented was described. PI selection procedures also include consideration and specification of the time base represented by the PI data. In Section 5.4 the necessity to carry out a PI assessment over prolonged periods was highlighted. This is necessary to build up a knowledge base and also in order to have enough information to make meaningful and robust comparisons for benchmarking. In this manual, the normal data reporting period is one year. However, it may be necessary to evaluate performance over shorter timescales, particularly for internal purposes and also because annually reported data may hide important data variability during a one year period. Where data are highly variable and reported only annually this may necessitate the designation of only limited reliability (e.g. Band C, Section 6.2), depending upon the results from statistical analysis of the data set that makes up the reported value.

In this manual and accompanying software the user has freedom to specify the time period over which data comprising each of the variables have been collected (Section 7.1). The time period is defined in the manual as the *assessment period.* Users of the manual and software should be aware that there might be a number of different time periods used within the data sets. It is also important to appreciate that the behaviour of most of the variables used to define the PIs is not uniform during a year, due to random or seasonal effects, or to other undertaking intervention activities.

It is strongly recommended that only those PIs based on variables with a one-year data record should be used for benchmarking performance with other undertakings. PIs based on data collected over shorter assessment periods should only be used for internal comparisons and short-term control purposes (e.g. comparisons of PIs between successive months or quarters).

7. PERFORMANCE INDICATORS

7.1 The PI framework

This Section contains the full listing of the WW performance indicators for wastewater services. Each undertaking should select the relevant subset of PIs depending upon the assessment objectives as described in Section 5. Where the wastewater undertaking is also responsible for services other than wastewater, only the core business component of wastewater should be considered for the assessment of the indicators here.

The PIs are each ratios between *variables*. The variables quantify the data for the various areas and items of interest. A variable is never a PI, however, it may provide context information. Hence some PIs are determined from ratios between variables, one or more of which may also be context information.

The WW PIs have been adapted from the core water supply manual PI list, with advice from industry discussion, workshops with a wide range of *stakeholders*, and widespread web discussions. This was to define how the wastewater PI system needed to differ from the PIs appropriate for water supply services. The added complexity of the wastewater system, the variability of the drivers and the different boundaries, constraints and impacts resulted in a new category of environmentally related PIs. The complexity also led to the need for a greater number of operationally related PIs and more detailed CI requirements. The initial draft list of wastewater PIs was refined by core group discussion and widespread dissemination and feedback from interested stakeholders, mainly via the World Wide Web.

The normal use of the wastewater PI system is for indicators to be reported using variables recorded with an annual frequency. It is recommended that the year be used as the reference assessment period as this provides a sufficient period for most PI data to be statistically consistent. However, many undertakings may need to track certain aspects of their performance more frequently using a limited number of key PIs. Hence, the PI system has been set up to accommodate user-defined individually specified PI assessment periods. Nonetheless, even where the variables comprising a PI are recorded for a period other than a year, the PI processing rule will ensure that the PI is reported to an equivalent base time period of one year. This is to avoid spurious comparisons between PIs with variables with non-commensurate time bases and to ensure unit reporting coherence. This adjustment when processing PIs is not intended to be an extrapolation from periods shorter than a year to represent a full year, but simply to provide a consistent time base. It is important to appreciate that the behaviour of most of the variables used to define the PIs is not uniform during a year, due to random or seasonal effects, or to other undertaking intervention activities.

When used for comparative purposes, it is essential to acknowledge those PIs that have been assessed from variables that have not been recorded annually. This is to avoid any bias or spurious misinterpretation of the results. It is strongly recommended that only those PIs based on variables with a one-year data record should be used for benchmarking performance with other undertakings. PIs based on data collected over shorter assessment periods should only be used for internal comparisons and short-term control purposes (e.g. comparisons of PIs between successive months or quarters).

In the main PI listings, which follow, advice is provided with each regarding the appropriate assessment period for the variables used to calculate the PI. There are four options:

- PI that may be assessed for periods shorter than one year;
- PI that may be assessed for periods shorter than one year, but it is recommended that it be used only where data for the variables have been collected for at least a year;
- PI that must only be assessed using annual data;
- PI recommended to be assessed using annual data for periods longer than one year

As described in Section 4.7, The PIs are structured into six groups: environmental, personnel, physical, operational, quality of service and economic and financial. The PIs may be classified into one of three levels of significance, according to the importance to the objective of the assessment and the context of application (Section 5.3). This distinction into significance levels also allows for a gradual, successive implementation of the full set of PIs.

Data acquisition needs for the PIs may need to be continuous throughout the assessment period, as variables may represent a cumulative value such as the volume of wastewater treated or the number of working accidents. Other variables may be more amenable to being obtained retrospectively at the end of the assessment period, at a fixed point in time. Such as total sewer length or number of pumping stations. In the latter case, it is essential that the undertaking have a reliable method for obtaining this data, or a reliable database.

In application many of the PIs are ratios of commensurate units, giving a dimensionless result. Some of these PIs are expressed as percentages by multiplication by 100. Where the percentage could lead to values exceeding 100% the PI is not expressed as a percentage (-).

There are different types of PI in the listings below. Whilst most of the PIs illustrate aspects of the wastewater system and service provision that are under the direct control of the undertaker, some of the PIs are neutral, i.e. they represent indicators that are not directly within the control of the undertaker. Although not necessarily within the direct control of the undertaker, these PIs still provide information vital to the delivery of the service.

Examples of neutral PIs include those that are sediment and solids related:

- *wEn12* - sediments from sewers;
- *wEn13* - sediments from ancillaries;
- *wEn15* - sediments from on-site systems;
- *wEn14* - solid waste from screens and grit.

Each of these is highly dependent on the characteristics of the catchment and of the sewer system and may not be directly dependent on the undertaking performance. These PIs could potentially be used strategically to improve services, by for example, encouraging highway drainage Authorities to control road runoff more effectively, so reducing the amount of sediments in sewers.

7.2 Environmental indicators (wEn)

Environmental indicators evaluate the performance of the undertaking regarding environmental impacts, including the compliance with wastewater discharge standards, intermittent overflow discharges and final disposal of solid wastes (sludge, sediments and screenings).

SUBGROUP	Code	Indicator (unit)	Concept / Processing rule / Comments
WASTEWATER	*wEn1*	WWTP compliance with discharge consents (%/year)	(*population equivalent* that is served by *wastewater treatment plants* complying with discharge consents / *population equivalent served by wastewater treatment plants* managed by the undertaking x 100, at the reference date *wEn1 = wA1 / wE5 x 100* *Discharge consents refer to the effluent quality standards that apply. The compliance is assessed regarding the loads or concentrations and their potential environmental impacts.* *This PI may be assessed for periods shorter than one year, but it is recommended that it be used only where data for the variables have been collected for at least a year. Where it has been used for shorter time periods, special consideration is required when used for comparisons, either internal or external to the undertaking. See section 7.1.*
	wEn2	Wastewater reuse (%)	Volume of reused treated wastewater / volume of wastewater treated by the undertaking x 100, during the assessment period *wEn2 = wA12 / wA2 x 100* *This indicator may be assessed for periods shorter than one year, but special consideration is required when used for comparisons, either internal or external to the undertaking. See section 7.1.*
	wEn3	*intermittent overflow discharge* frequency (No./overflow device/year)	(Number of *overflow* discharges that occurred during the assessment period x 365 / assessment period) / number of overflow devices at the reference date *wEn3 = (wA24 x 365 / wH1) / wC19* *Note that " x 365 / assessment period" is a unit conversion expression and is not intended to be considered as extrapolation.* *This PI may be assessed for periods shorter than one year, but it is recommended that it be used only where data for the variables have been collected for at least a year. Where it has been used for shorter time periods, special consideration is required when used for comparisons, either internal or external to the undertaking. See section 7.1.*
	wEn4	*intermittent overflow discharge* volume (m^3/overflow device/year)	(Total volume of *overflow* discharges that occurred during the assessment period x 365 / assessment period) / number of overflow devices at the reference date *wEn4 = (wA25 x 365 / wH1) / wC19* *Note that " x 365 / assessment period" is a unit conversion expression and is not intended to be considered as extrapolation.* *This PI may be assessed for periods shorter than one year, but it is recommended that it be used only where data for the variables have been collected for at least a year. Where it has been used for shorter time periods, special consideration is required when used for comparisons, either internal or external to the undertaking. See section 7.1.*
	wEn5	*intermittent overflow discharge* related to rainfall (%/year)	Total volume of *overflow* discharges / rainfall volume x 100, during the assessment period *wEn5 = wA25 / wA26 x 100* *This PI may be assessed for periods shorter than one year, but it is recommended that it be used only where data for the variables have been collected for at least a year, in order to allow for comparisons with wEn4.*

SUBGROUP	*Code*	Indicator *(unit)*	Concept / *Processing rule* / *Comments*
SOLID RESIDUES	*wEn6*	Sludge production in WWTP *(kg DS/p.e./year)*	(Dry weight of *sludge* produced in *wastewater treatment plants* managed by the undertaking during the assessment period x 365 / assessment period) / *population equivalent served by wastewater treatment plants* at the reference date x 1000 *wEn6 = (wA13 x 365 / wH1) / wE5 x 1000* *Note that " x 365 / assessment period" is a unit conversion expression and is not intended to be considered as extrapolation.* *This PI may be assessed for periods shorter than one year, but it is recommended that it be used only where data for the variables have been collected for at least a year. Where it has been used for shorter time periods, special consideration is required when used for comparisons, either internal or external to the undertaking. See section 7.1.*
	wEn7	Sludge utilisation *(%)*	Dry weight of *sludge utilised* / dry weight of sludge handled x 100, during the assessment period *wEn7 = wA15 / wA14 x 100* *Sludge utilisation includes all types of uses (agriculture, forest, products, materials, etc.).* *This indicator may be assessed for periods shorter than one year, but special consideration is required when used for comparisons, either internal or external to the undertaking. See section 7.1.*
	wEn8	Sludge disposal *(%)*	Dry weight of *sludge disposed* / dry weight of sludge handled x 100, during the assessment period *wEn8 = wA16 / wA14 x 100* *Sludge disposal includes sludge going to landfill and residues from sludge utilisation ation processes.* *This indicator may be assessed for periods shorter than one year, but special consideration is required when used for comparisons, either internal or external to the undertaking. See section 7.1.*
	wEn9	- sludge going to landfill *(%)*	Dry weight of *sludge* going to landfill / dry weight of sludge handled x 100, during the assessment period *wEn9 = wA17 / wA14 x 100* *This indicator may be assessed for periods shorter than one year, but special consideration is required when used for comparisons, either internal or external to the undertaking. See section 7.1.*
	wEn10	- sludge thermally processed *(%)*	Dry weight of *sludge* thermally processed / dry weight of sludge handled x 100, during the assessment period *wEn10 = wA18 / wA14 x 100* *This indicator may be assessed for periods shorter than one year, but special consideration is required when used for comparisons, either internal or external to the undertaking. See section 7.1.*
	wEn11	- other sludge disposal *(%)*	Dry weight of *sludge disposed* by other means than in wEn9 and wEn10 / dry weight of sludge handled x 100, during the assessment period *wEn11 = wA19 / wA14 x 100* *This indicator may be assessed for periods shorter than one year, but special consideration is required when used for comparisons, either internal or external to the undertaking. See section 7.1.*

SUBGROUP *Code*	Indicator *(unit)*	Concept *Processing rule* *Comments*
wEn12	Sediments from sewers *(ton/km sewer/year)*	(Drained weight of *sediments* removed from sewers during the assessment period x 365 / assessment period) / total sewer length at the reference date *wEn12 = (wA20 x 365 / wH1) / wC1* *Note that " x 365 / assessment period" is a unit conversion expression and is not intended to be considered as extrapolation.* *This PI may be assessed for periods shorter than one year, but it is recommended that it be used only where data for the variables have been collected for at least a year. Where it has been used for shorter time periods, special consideration is required when used for comparisons, either internal or external to the undertaking. See section 7.1.*
wEn13	Sediments from ancillaries *(ton/km sewer/year)*	(Drained weight of *sediments* removed from *ancillaries* during the assessment period x 365 / assessment period) / total sewer length at the reference date *wEn13 = (wA21 x 365 / wH1) / wC1* *Note that " x 365 / assessment period" is a unit conversion expression and is not intended to be considered as extrapolation.* *This PI may be assessed for periods shorter than one year, but it is recommended that it be used only where data for the variables have been collected for at least a year. Where it has been used for shorter time periods, special consideration is required when used for comparisons, either internal or external to the undertaking. See section 7.1.*
wEn14	Solid waste from screens and grit *(ton/km sewer/year)*	(Drained weight of screenings and grit removed from *wastewater treatment plants* and *pumping stations* during the assessment period x 365 / assessment period) / total sewer length at the reference date *wEn14 = (wA22 x 365 / wH1) / wC1* *Note that " x 365 / assessment period" is a unit conversion expression and is not intended to be considered as extrapolation.* *This PI may be assessed for periods shorter than one year, but it is recommended that it be used only where data for the variables have been collected for at least a year. Where it has been used for shorter time periods, special consideration is required when used for comparisons, either internal or external to the undertaking. See section 7.1.*
wEn15	Sediments from on-site systems *(ton/p.e./year)*	(Drained weight of *sediments* removed from *on-site systems* during the assessment period x 365 / assessment period) / *population equivalent served by on-site systems* that are the responsibility of the undertaking at the reference date *wEn15 = (wA23 x 365 / wH1) / wE6* *Note that " x 365 / assessment period" is a unit conversion expression and is not intended to be considered as extrapolation.* *This PI may be assessed for periods shorter than one year, but it is recommended that it be used only where data for the variables have been collected for at least a year. Where it has been used for shorter time periods, special consideration is required when used for comparisons, either internal or external to the undertaking. See section 7.1.*

7.3 Personnel indicators (wPe)

Personnel indicators assess efficiency and effectiveness of the wastewater undertaking personnel, considering functions, activities and qualifications. Matters like training, health and safety and absenteeism are also taken into account. Correct interpretation of these PI entail a cross-reference to outsourcing data.

Employees include every person who works for the undertaking in return for a wage.

SUBGROUP	*Code*	Indicator *(unit)*	Concept / *Processing rule* / *Comments*
TOTAL PERSONNEL	*wPe1*	Personnel in WWT per population equivalent *(No./ 1000 p.e.)*	Number of full time equivalent employees working on wastewater treatment / *population equivalent* served by wastewater treatment managed by the undertaking x 1000, at the reference date *wPe1 = wB2 / (wE5 + wE6) x 1000* *WWT refers to both end of pipe treatment plants and on-site systems that are the responsibility of the undertaking.*
	wPe2	Personnel working on SE per length of sewer *(No./100 km sewer)*	Number of full time equivalent employees working on *sewer system* / total sewer length x 100, at the reference date *wPe2 = wB3 / wC1 x 100*
PERSONNEL PER MAIN FUNCTION	*wPe3*	General management personnel *(%)*	Number of full time equivalent employees dedicated to directorate, central administration, strategic planning, public relations and other *stakeholder* relations, legal affairs, internal audits, environmental management, general computing support and new business development activities / number of full time equivalent employees of the wastewater undertaking x 100, at the reference date *wPe3 = wB4 / wB1 x 100*
	wPe4	Human resources management personnel *(%)*	Number of full time equivalent employees dedicated to personnel administration, education and training, occupational safety and medicine services and social activities / number of full time equivalent employees of the wastewater undertaking x 100, at the reference date *wPe4 = wB5 / wB1 x 100*
	wPe5	Financial and commercial personnel *(%)*	Number of full time equivalent employees dedicated to economic and financial planning, economic administration, economic controlling, purchasing and material management activities / number of full time equivalent employees of the wastewater undertaking x 100, at the reference date *wPe5 = wB6 / wB1 x 100*
	wPe6	Customer service personnel *(%)*	Number of full time equivalent employees dedicated to accounting and control and to customer relations and management activities / number of full time equivalent employees of the wastewater undertaking x 100, at the reference date *wPe6 = wB7 / wB1 x 100*

SUBGROUP	*Code*	Indicator *(unit)*	Concept *Processing rule* *Comments*
	wPe7	Technical services personnel *(%)*	Number of full time equivalent employees dedicated to planning, design, construction, operation, maintenance and *repair* activities / number of full time equivalent employees of the wastewater undertaking x 100 *wPe7 = wB8 / wB1 x 100*
	wPe8	- planning, design and construction personnel *(%)*	Number of full time equivalent employees dedicated to planning, design and construction of SE and wastewater treatment facilities / number of full time equivalent employees of the wastewater undertaking x 100, at the reference date *wPe8 = wB11 / wB1 x 100*
	wPe9	- operation and maintenance personnel *(%)*	Number of full time equivalent employees dedicated to operation, maintenance and *repair* of SE and wastewater treatment facilities / number of full time equivalent employees of the wastewater undertaking x 100, at the reference date *wPe9 = wB12 / wB1 x 100*
TECHNICAL PERSONNEL PER ACTIVITY	*wPe10*	Technical WWT personnel *(No./1000 p.e.)*	Number of full time equivalent employees working on WWT planning, design, construction, operation, maintenance and *repair* activities / *population equivalent* served by wastewater treatment managed by the undertaking x 1000, at the reference date *wPe10 = wB9 / (wE5 + wE6) x 1000* *In this PI, WWT refers to both end of pipe treatment plants and on-site systems that are the responsibility of the undertaking.* *When the undertaking assets include laboratories, the wastewater quality sampling and testing personnel shall be accounted in the performance indicator wPe12.*
	wPe11	Technical SE personnel *(No./100 km sewer)*	Number of full time equivalent employees working on SE planning, design, operation, maintenance and *repair* / total sewer length x 100, at the reference date *wPe11 = wB10 / wC1 x 100* *When the undertaking assets include laboratories, the wastewater quality sampling and testing personnel shall be accounted in the performance indicator wPe12.*
	wPe12	Wastewater quality monitoring personnel *(No./(1000 tests/year))*	Number of full time equivalent employees working on wastewater quality sampling and testing / (total number of wastewater quality tests carried out by the undertaking laboratories x 365 / assessment period) x 1000 *wPe12 = wB13 / (wD47 x 365 / wH1) x 1000* *Note that " x 365 / assessment period" is a unit conversion expression and is not intended to be considered as extrapolation.* *This indicator may be assessed for periods shorter than one year, but special consideration is required when used for comparisons, either internal or external to the undertaking. See section 7.1.* *Applicable only when the wastewater undertaking performs quality tests in its own laboratories.*

SUBGROUP	Code	Indicator (unit)	Concept / Processing rule / Comments
	wPe13	Support services personnel (%)	Number of full time equivalent employees working on support services / total number of full time equivalent employees working on technical services x 100, at the reference date *wPe13 = wB14 / wB8 x 100*
PERSONNEL QUALIFICATION	wPe14	University degree personnel (%)	Number of full time equivalent wastewater undertaking employees with university degree / number of full time equivalent employees of the wastewater undertaking x 100, at the reference date *wPe14 = wB15 / wB1 x 100*
	wPe15	Basic education personnel (%)	Number of full time equivalent wastewater undertaking employees with basic education but without university degree / number of full time equivalent employees of the wastewater undertaking x 100, at the reference date *wPe15 = wB16 / wB1 x 100*
	wPe16	Other qualification personnel (%)	Total number of full-time equivalent employees without basic education / number of full time equivalent employees of the wastewater undertaking x 100, at the reference date or 100 – (university degree personnel + basic education personnel), at the reference date *wPe16 = wB17 / wB1 x 100* *or wPe16 = 100 - (wPe14+wPe15)*
PERSONNEL TRAINING	wPe17	Total training of personnel (hours/employee/year)	(Number of training hours for wastewater personnel during the assessment period x 365 / assessment period) / number of full time equivalent employees of the wastewater undertaking at the reference date *wPe17 = (wB20 x 365 / wH1) / wB1* *Note that " x 365 / assessment period" is a unit conversion expression and is not intended to be considered as extrapolation.* *This PI may be assessed for periods shorter than one year, but it is recommended that it be used only where data for the variables have been collected for at least a year. Where it has been used for shorter time periods, special consideration is required when used for comparisons, either internal or external to the undertaking. See section 7.1.*
PERSONNEL VACCINATION AND SAFETY	wPe18	Vaccination (%)	Number of full time equivalent wastewater undertaking employees with up to date vaccination certification for *sewage* related diseases / operation and maintenance personnel x 100, at the reference date *wPe18 = wB19 / wB12 x 100*
	wPe19	Authorised confined spaces training (%)	Number of full time equivalent wastewater undertaking employees with appropriate authorised confined spaces training / operation and maintenance personnel x 100, at the reference date *wPe19 = wB18 / wB12 x 100*

SUBGROUP	*Code*	Indicator *(unit)*	Concept / *Processing rule* / *Comments*
	wPe20	Working accidents *(No./100 employees/year)*	(Number of working accidents requiring medical care that occurred with personnel during the assessment period x 365 / assessment period) / number of full time equivalent employees of the wastewater undertaking at the reference date x 100 *wPe20 = (wB21 x 365 / wH1) / wB1 x 100* *Note that " x 365 / assessment period" is a unit conversion expression and is not intended to be considered as extrapolation.* *This PI may be assessed for periods shorter than one year, but it is recommended that it be used only where data for the variables have been collected for at least a year. Where it has been used for shorter time periods, special consideration is required when used for comparisons, either internal or external to the undertaking. See section 7.1.*
	wPe21	- personnel working fatalities *(No./100 employees/year)*	(Number of working accidents resulting in death or permanent illness during the assessment period x 365 / assessment period) / number of full time equivalent employees of the wastewater undertaking at the reference date x 100 *wPe21 = (wB22 x 365 / wH1) / wB1 x 100* *Note that " x 365 / assessment period" is a unit conversion expression and is not intended to be considered as extrapolation.* *This PI may be assessed for periods shorter than one year, but it is recommended that it be used only where data for the variables have been collected for at least a year. Where it has been used for shorter time periods, special consideration is required when used for comparisons, either internal or external to the undertaking. See section 7.1.*
ABSENTEEISM	*wPe22*	Absenteeism *(days/100 employees/year)*	(Number of days of personnel absenteeism during the assessment period x 365 / assessment period) / number of full time equivalent employees of the wastewater undertaking at the reference date x 100 *wPe22 = (wB23 x 365 / wH1) /wB1* *Note that " x 365 / assessment period" is a unit conversion expression and is not intended to be considered as extrapolation.* *This PI may be assessed for periods shorter than one year, but it is recommended that it be used only where data for the variables have been collected for at least a year. Where it has been used for shorter time periods, special consideration is required when used for comparisons, either internal or external to the undertaking. See section 7.1.*

SUBGROUP	Code	Indicator (unit)	Concept / Processing rule / Comments
	wPe23	- absenteeism due to accidents or illness at work *(days/100 employees/year)*	(Number of days of personnel absenteeism due to accidents or illness during the assessment period x 365 / assessment period) / number of full time equivalent employees of the wastewater undertaking at the reference date x 100 *wPe23 = (wB24 x 365 / wH1) / wB1* *Note that " x 365 / assessment period" is a unit conversion expression and is not intended to be considered as extrapolation.* *This PI may be assessed for periods shorter than one year, but it is recommended that it be used only where data for the variables have been collected for at least a year. Where it has been used for shorter time periods, special consideration is required when used for comparisons, either internal or external to the undertaking. See section 7.1.*
	wPe24	- absenteeism due to other reasons *(days/100 employees/year)*	(Number of days of personnel absenteeism that were not due to working accidents or illness during the assessment period x 365 / assessment period) / number of full time equivalent employees of the wastewater undertaking at the reference date x 100 *wPe24 = (wB25 x 365 / wH1) / wB1* *or wPe24 = wPe22 - wPe23* *Note that " x 365 / assessment period" is a unit conversion expression and is not intended to be considered as extrapolation.* *This PI may be assessed for periods shorter than one year, but it is recommended that it be used only where data for the variables have been collected for at least a year. Where it has been used for shorter time periods, special consideration is required when used for comparisons, either internal or external to the undertaking. See section 7.1.*
OVERTIME WORK	*wPe25*	Overtime work *(%)*	Overtime work / normal work x 100, during the assessment period *wPe25 = wB27 / wB26 x 100* *This indicator is meant for skilled and unskilled workers; too much overtime work can indicate inefficient use of manpower, lack of personnel or inefficient preventive maintenance.* *This indicator may be assessed for periods shorter than one year, but special consideration is required when used for comparisons, either internal or external to the undertaking. See section 7.1.*

7.4 Physical indicators (wPh)

Physical PIs aim to evaluate if wastewater treatment and sewerage assets still have enough capacity (headroom) to operate correctly and safely, assuring that their service targets can be attained. The utilisation of preliminary, primary, secondary and tertiary treatment is considered as well as the degree of surcharging in the sewers. Pumping capacity utilisation and automation and the degree of control are also included .

SUBGROUP	*Code*	Indicator (unit)	Concept / *Processing rule* / *Comments*
WASTEWATER TREATMENT	*wPh1*	Preliminary treatment utilisation (%)	Maximum sum of daily volume of treated wastewater in all *preliminary wastewater treatment plants* / Sum, for all preliminary wastewater treatment plants, of the daily maximum treatment capacities x 100, during the assessment period *wPh1 = wA4 / wC12 x 100* *This indicator may be assessed for periods shorter than one year, but special consideration is required when used for comparisons, either internal or external to the undertaking. See section 7.1.*
	wPh2	Primary treatment utilisation (%)	Maximum sum of daily volume of treated wastewater in *primary wastewater treatment plants* / Sum, for all primary wastewater treatment plants, of the daily maximum treatment capacities x 100, during the assessment period *wPh2 = wA6 / wC13 x 100* *This indicator may be assessed for periods shorter than one year, but special consideration is required when used for comparisons, either internal or external to the undertaking. See section 7.1.*
	wPh3	Secondary treatment utilisation (%)	Maximum sum of daily volume of treated wastewater in *secondary wastewater treatment plants* / Sum, for all secondary wastewater treatment plants, of the daily maximum treatment capacities x 100, during the assessment period *wPh3 = wA8 / wC14 x 100* *This indicator may be assessed for periods shorter than one year, but special consideration is required when used for comparisons, either internal or external to the undertaking. See section 7.1.*
	wPh4	Tertiary treatment utilisation (%)	Maximum sum of daily volume of treated wastewater in *tertiary wastewater treatment plants* / Sum, for all tertiary wastewater treatment plants, of the daily maximum treatment capacities x 100, during the assessment period *wPh4 = wA10 / wC15 x 100* *This indicator may be assessed for periods shorter than one year, but special consideration is required when used for comparisons, either internal or external to the undertaking. See section 7.1.*
SEWERS	*wPh5*	Surcharging in gravity sewers in dry weather (%)	Length of gravity sewer where *surcharging* has occurred in *dry weather* during the assessment period / total sewer length at the reference date x 100 *wPh5 = wC2 / wC1 x 100* *This PI may be assessed for periods shorter than one year, but it is recommended that it be used only where data for the variables have been collected for at least a year. Where it has been used for shorter time periods, special consideration is required when used for comparisons, either internal or external to the undertaking. See section 7.1.*

SUBGROUP	*Code*	Indicator *(unit)*	Concept / *Comments* / *Processing rule*
	wPh6	Surcharging in gravity sewers in wet weather *(%)*	Length of gravity sewers where *surcharging* has occurred in *wet weather* during the assessment period / total sewer length at the reference date x 100 *wPh6 = wC3 / wC1 x 100* *This information may be obtained either by monitoring or by hydraulic modelling using real rainfall data.* *This PI may be assessed for periods shorter than one year, but it is recommended that it be used only where data for the variables have been collected for at least a year. Where it has been used for shorter time periods, special consideration is required when used for comparisons, either internal or external to the undertaking. See section 7.1.*
	wPh7	High sewer surcharging *(%)*	Length of sewer where a high degree of *surcharging* has occurred in *wet weather* during the assessment period / total sewer length at the reference date x 100 *wPh7 = wC4 / wC1 x 100* *High degree of surcharging means water level at least 0.5 m above the pipe crown.* *This PI may be assessed for periods shorter than one year, but it is recommended that it be used only where data for the variables have been collected for at least a year. Where it has been used for shorter time periods, special consideration is required when used for comparisons, either internal or external to the undertaking. See section 7.1.*
PUMPING HEADROOM	*wPh8*	Pump power utilised in SE *(%)*	$\Sigma_{\text{for all pumps installed in SE}}$(pump nominal power x pump working hours during the assessment period) / (total pump nominal power installed in the *sewer system* x assessment period x 24) x 100 *wPh8 = wD15 / (wC10 x wH1 x 24) x 100* *Pump power installed does not include standby pumps (those used for substitution).* *This indicator may be assessed for periods shorter than one year, but special consideration is required when used for comparisons, either internal or external to the undertaking. See section 7.1.*
	wPh9	Pump power utilised in WWTP *(%)*	$\Sigma_{\text{for all pumps installed in WWTP}}$(pump nominal power x pump working hours during the assessment period) / (total pump nominal power installed in *wastewater treatment plants* x assessment period x 24) x 100 *wPh9 = wD14 / (wC5 x wH1 x 24) x 100* *Wastewater treatment plant inlet pumps are to be included.* *This indicator may be assessed for periods shorter than one year, but special consideration is required when used for comparisons, either internal or external to the undertaking. See section 7.1.*

SUBGROUP	Code	Indicator (unit)	Concept / Processing rule / Comments
	wPh10	Sewer system pump headroom (%)	Number of *pumping stations* in the *sewer system* where pumps operated more than 75% of the time during the assessment period / number of sewer system pumping stations at the reference date x 100 *wPh10 = wC11 / wC9 x 100* *Pumps in wastewater treatment plants shall not be accounted for.* *This indicator may be assessed for periods shorter than one year, but special consideration is required when used for comparisons, either internal or external to the undertaking. See section 7.1.*
AUTOMATION AND CONTROL	wPh11	Automation degree (%)	Number of automated control units / number of control units x 100, at the reference date *wPh11 = wC17 / wC16 x 100*
	wPh12	Remote control degree (%)	Number of remote control units / number of control units x 100, at the reference date *wPh12 = wC18 / wC16 x 100*

7.5 Operational indicators (wOp)

In this group, PIs are intended to assess the performance of the undertaking as regards operation and maintenance activities. The areas to be assessed include sewers, ancillaries, pumps and pumping station inspection and maintenance, equipment calibration, electrical equipment inspection, energy consumption, sewer and pump rehabilitation, inflow/infiltration/exfiltration, failures, wastewater and sludge quality monitoring, vehicle availability and safety equipment.

SUBGROUP	Code	Indicator (unit)	Concept / Processing rule / Comments
SEWER INSPECTION AND MAINTENANCE	wOp1	Sewer inspection (%/year)	(Length of sewers *inspected* during the assessment period x 365 / assessment period) / total sewer length at the reference date x 100 *wOp1 = (wD1 x 365 / wH1) / wC1 x 100* *On-line inspection by remote surveillance and TV inspection shall be included. Also inspections to investigate blockages are included.* *Note that " x 365 / assessment period" is a unit conversion expression and is not intended to be considered as extrapolation.* *This PI may be assessed for periods shorter than one year, but it is recommended that it be used only where data for the variables have been collected for at least a year. Where it has been used for shorter time periods, special consideration is required when used for comparisons, either internal or external to the undertaking. See section 7.1.*

SUBGROUP	Code	Indicator (unit)	Concept / Processing rule / Comments
	wOp2	Sewer cleaning *(%/year)*	(Length of sewers *cleaned* during the assessment period x 365 / assessment period) / total sewer length at the reference date x 100 *wOp2 = (wD2 x 365 / wH1) / wC1 x 100* *Sewer cleaning refers to actions under a proactive management strategy. Root cutting shall be included. Curative cleaning of blockages is reflected in wOp34.* *Note that " x 365 / assessment period" is a unit conversion expression and is not intended to be considered as extrapolation.* *This PI may be assessed for periods shorter than one year, but it is recommended that it be used only where data for the variables have been collected for at least a year. Where it has been used for shorter time periods, special consideration is required when used for comparisons, either internal or external to the undertaking. See section 7.1.*
	wOp3	Manhole chamber inspection *(-/year)*	(Number of manhole chambers *inspected* during the assessment period x 365 / assessment period) / total number of manhole chambers at the reference date *wOp3 = (wD3 x 365 / wH1) / wC21* *Note that " x 365 / assessment period" is a unit conversion expression and is not intended to be considered as extrapolation.* *This PI may be assessed for periods shorter than one year, but it is recommended that it be used only where data for the variables have been collected for at least a year. Where it has been used for shorter time periods, special consideration is required when used for comparisons, either internal or external to the undertaking. See section 7.1.*
	wOp4	Gully pot inspection *(-/year)*	(Number of *gully pots inspected* during the assessment period x 365 / assessment period) / total number of gully pots at the reference date *wOp4 = (wD4 x 365 / wH1) / wC22* *Gully pot inspection refers to actions under proactive management strategy.* *If this activity is not the responsibility of the wastewater undertaking, this PI is not applicable. However, this information is important to help in understanding wQS14.* *Note that " x 365 / assessment period" is a unit conversion expression and is not intended to be considered as extrapolation.* *This PI may be assessed for periods shorter than one year, but it is recommended that it be used only where data for the variables have been collected for at least a year. Where it has been used for shorter time periods, special consideration is required when used for comparisons, either internal or external to the undertaking. See section 7.1.*

SUBGROUP	Code	Indicator (unit)	Concept / Processing rule / Comments
	wOp5	Gully pot cleaning *(-/year)*	(Number of *gully pots cleaned* during the assessment period x 365 / assessment period) / total number of gully pots at the reference date *wOp5 = (wD5 x 365 / wH1) / wC22* *Gully pot cleaning refers to actions under proactive management strategy. If this activity is not the responsibility of the wastewater undertaking, this PI is not applicable. However, this information is important to help in understanding wQS14.* *Note that " x 365 / assessment period" is a unit conversion expression and is not intended to be considered as extrapolation.* *This PI may be assessed for periods shorter than one year, but it is recommended that it be used only where data for the variables have been collected for at least a year. Where it has been used for shorter time periods, special consideration is required when used for comparisons, either internal or external to the undertaking. See section 7.1.*
TANKS AND CSOs INSPECTION AND MAINTENANCE	*wOp6*	Tanks and CSOs inspection frequency *(No./tank and CSO/year)*	(Number of *inspections* of tanks and *CSOs* carried out during the assessment period x 365 / assessment period) / total number of tanks and CSOs at the reference date *wOp6 = (wD6 x 365 / wH1) / (wC20+wC23)* *Note that " x 365 / assessment period" is a unit conversion expression and is not intended to be considered as extrapolation.* *This PI may be assessed for periods shorter than one year but the resulting values can be misleading. Thus it is recommended to be assessed using annual data for periods of more than one year. Where it has been used for shorter time periods, special attention shall be paid when this PI is used for comparisons, either internal or external to the undertaking. See section 7.1.*
	wOp7	Tank and CSOs inspection volume *(-/year)*	(Volume of tanks and *CSOs inspected* during the assessment period x 365 / assessment period) / total volume of tanks and CSOs at the reference date *wOp7 = (wD7 x 365 / wH1) / wC24* *Only CSOs with storage capacity are included.* *Note that " x 365 / assessment period" is a unit conversion expression and is not intended to be considered as extrapolation.* *This PI may be assessed for periods shorter than one year but the resulting values can be misleading. Thus it is recommended to be assessed using annual data for periods of more than one year. Where it has been used for shorter time periods, special attention shall be paid when this PI is used for comparisons, either internal or external to the undertaking. See section 7.1.*

SUBGROUP	Code	Indicator (unit)	Concept / Processing rule / Comments
	wOp8	Tank and CSOs cleaning (-/year)	(Volume of tanks and *CSOs cleaned* during the assessment period x 365 / assessment period) / total volume of tanks and CSOs at the reference date *wOp8 = (wD8 x 365 / wH1) / wC24* *Only CSOs with storage capacity are included.* *Note that " x 365 / assessment period" is a unit conversion expression and is not intended to be considered as extrapolation.* *This PI may be assessed for periods shorter than one year but the resulting values can be misleading. Thus it is recommended to be assessed using annual data for periods of more than one year. Where it has been used for shorter time periods, special attention shall be paid when this PI is used for comparisons, either internal or external to the undertaking. See section 7.1.*
	wOp9	Screen inspection (-/year)	(Number of *screens* associated with tanks and *CSOs inspected* during the assessment period x 365 / assessment period) / number of screens associated with tanks and CSOs at the reference date *wOp9 = (wD9 x 365 / wH1) / wC25* *Note that " x 365 / assessment period" is a unit conversion expression and is not intended to be considered as extrapolation.* *This PI may be assessed for periods shorter than one year, but it is recommended that it be used only where data for the variables have been collected for at least a year. Where it has been used for shorter time periods, special consideration is required when used for comparisons, either internal or external to the undertaking. See section 7.1.*
PUMPS AND PUMPING STATIONS INSPECTION	*wOp10*	Pumping station inspection frequency (-/year)	(Number of *inspections* to *pumping stations* carried out during the assessment period x 365 / assessment period) / total number of pumping stations at the reference date *wOp10 = (wD10 x 365 / wH1) / wC6* *wastewater treatment plant pumps should not be included.* *Note that " x 365 / assessment period" is a unit conversion expression and is not intended to be considered as extrapolation.* *This indicator may be assessed for periods shorter than one year, but special consideration is required when used for comparisons, either internal or external to the undertaking. See section 7.1.*
	wOp11	Pump inspection by power (-/year)	(Total nominal power of pumps and related *ancillaries* subject to *inspection* during the assessment period x 365 / assessment period) / total nominal power of pumps at the reference date *wOp11 = (wD11 x 365 / wH1) / wC10* *wastewater treatment plants pumps should not be included.* *Note that " x 365 / assessment period" is a unit conversion expression and is not intended to be considered as extrapolation.* *This indicator may be assessed for periods shorter than one year, but special consideration is required when used for comparisons, either internal or external to the undertaking. See section 7.1.*

SUBGROUP	*Code*	Indicator *(unit)*	Concept *Processing rule* *Comments*
EQUIPMENT CALIBRATION	*wOp12*	SE flow meters calibration *(-/year)*	(Number of *calibrations* carried out for flow meters permanently installed in the *sewer system* during the assessment period x 365 / assessment period) / number of permanently installed flowmeters in the sewer system at the reference date *wOp12 = (wD20 x 365 / wH1) / wC27* *Note that " x 365 / assessment period" is a unit conversion expression and is not intended to be considered as extrapolation.* *This indicator may be assessed for periods shorter than one year, but special consideration is required when used for comparisons, either internal or external to the undertaking. See section 7.1.*
	wOp13	WWTP flow meters calibration *(-/year)*	(Number of *calibrations* carried out during the assessment period for flow meters permanently installed in *wastewater treatment plants* x 365 / assessment period) / number of permanently installed flowmeters in wastewater treatment plants at the reference date *wOp13 = (wD19 x 365 / wH1) / wC26* *Note that " x 365 / assessment period" is a unit conversion expression and is not intended to be considered as extrapolation.* *This indicator may be assessed for periods shorter than one year, but special consideration is required when used for comparisons, either internal or external to the undertaking. See section 7.1.*
	wOp14	Wastewater quality monitoring equipment calibration *(-/year)*	(Number of permanent automatic wastewater quality monitoring instrument *calibrations* carried out in *wastewater treatment plants* during the assessment period x 365 / assessment period) / number of wastewater quality instruments installed permanently in wastewater treatment plants at the reference date *wOp14 = (wD21 x 365 / wH1) / wC33* *Note that " x 365 / assessment period" is a unit conversion expression and is not intended to be considered as extrapolation.* *This indicator may be assessed for periods shorter than one year, but special consideration is required when used for comparisons, either internal or external to the undertaking. See section 7.1.*
ELECTRICAL AND SIGNAL TRANSMISSION EQUIPMENT INSPECTION	*wOp15*	Emergency power system inspection *(-/year)*	(Sum of the nominal power of the emergency power systems *inspected* during the assessment period x 365 / assessment period) / total nominal power of the emergency power systems at the reference date x 100 *wOp15 = (wD22 x 365 / wH1) / wC30 x 100* *Note that " x 365 / assessment period" is a unit conversion expression and is not intended to be considered as extrapolation.* *This indicator may be assessed for periods shorter than one year, but special consideration is required when used for comparisons, either internal or external to the undertaking. See section 7.1.*

SUBGROUP	Code	Indicator (unit)	Concept / Processing rule / Comments
	wOp16	Signal transmission equipment inspection *(-/year)*	(Number of signal transmission units *inspected* during the assessment period x 365 / assessment period) / total number of signal transmission units at the reference date *wOp16 = wD23 / wC31 x 100* *Note that " x 365 / assessment period" is a unit conversion expression and is not intended to be considered as extrapolation.* *This indicator may be assessed for periods shorter than one year, but special consideration is required when used for comparisons, either internal or external to the undertaking. See section 7.1.*
	wOp17	Electrical switchgear equipment inspection *(-/year)*	(Number of electrical switchgear units *inspected* during the assessment period x 365 / assessment period) / total number of electrical switchgear units at the reference date *wOp17 = (wD24 x 365 / wH1) / wC32 x 100* *Note that " x 365 / assessment period" is a unit conversion expression and is not intended to be considered as extrapolation.* *This indicator may be assessed for periods shorter than one year, but special consideration is required when used for comparisons, either internal or external to the undertaking. See section 7.1.*
ENERGY CONSUMPTION	*wOp18*	WWT energy consumption *(kWh/p.e./year)*	(Energy consumed by wastewater treatment facilities during the assessment period x 365 / assessment period) / *population equivalent* served by *wastewater treatment plants* managed by the undertaking at the reference date *wOp18 = (wD13 x 365 / wH1) / wE5* *WWT include on-site systems. Thermal processing of sludge should not be included.* *Note that " x 365 / assessment period" is a unit conversion expression and is not intended to be considered as extrapolation.* *This indicator may be assessed for periods shorter than one year, but special consideration is required when used for comparisons, either internal or external to the undertaking. See section 7.1.*
	wOp19	WWT energy recovery from co-generation processes *(%)*	Energy produced by co-generation processes / energy consumption at wastewater treatment facilities x 100, during the assessment period *wOp19 = wD17 / wD13 x 100* *Note that " x 365 / assessment period" is a unit conversion expression and is not intended to be considered as extrapolation.* *This indicator may be assessed for periods shorter than one year, but special consideration is required when used for comparisons, either internal or external to the undertaking. See section 7.1.*
	wOp20	Standardised energy consumption *(kWh/m³/m)*	Energy consumed for pumping wastewater / Σ(wastewater volume pumped x pump head in meters), during the assessment period *wOp20 = wD12 / wD16* *Note that " x 365 / assessment period" is a unit conversion expression and is not intended to be considered as extrapolation.* *This indicator may be assessed for periods shorter than one year, but special consideration is required when used for comparisons, either internal or external to the undertaking. See section 7.1.*

SUBGROUP	Code	Indicator (unit)	Concept / Processing rule / Comments
SEWER SYSTEM REHABILITATION	*wOp21*	Sewer rehabilitation *(%/year)*	(Length of defective sewers *rehabilitated* during the assessment period x 365 / assessment period) / total sewer length at the reference date x 100 *wOp21 = (wD25 x 365 / wH1) / wC1 x 100* *Note that " x 365 / assessment period" is a unit conversion expression and is not intended to be considered as extrapolation.* *This PI may be assessed for periods shorter than one year, but it is recommended that it be used only where data for the variables have been collected for at least a year. Where it has been used for shorter time periods, special consideration is required when used for comparisons, either internal or external to the undertaking. See section 7.1.*
	wOp22	- sewer renovation *(%/year)*	(Length of defective sewers *renovated* during the assessment period x 365 / assessment period) / total sewer length at the reference date x 100 *wOp22 = (wD26 x 365 / wH1) / wC1 x 100* *Note that " x 365 / assessment period" is a unit conversion expression and is not intended to be considered as extrapolation.* *This PI may be assessed for periods shorter than one year, but it is recommended that it be used only where data for the variables have been collected for at least a year. Where it has been used for shorter time periods, special consideration is required when used for comparisons, either internal or external to the undertaking. See section 7.1.*
	wOp23	- sewer replacement *(%/year)*	(Length of sewers *replaced* (or *renewed*) during the assessment period x 365 / assessment period) / total sewer length at the reference date x 100 *wOp23 = (wD27 x 365 / wH1) / wC1 x 100* *Note that " x 365 / assessment period" is a unit conversion expression and is not intended to be considered as extrapolation.* *This PI may be assessed for periods shorter than one year, but it is recommended that it be used only where data for the variables have been collected for at least a year. Where it has been used for shorter time periods, special consideration is required when used for comparisons, either internal or external to the undertaking. See section 7.1.*
	wOp24	- sewer and joints repair *(No./ 100 km sewer/year)*	(Number of *repairs* carried out in defective sewers and joints during the assessment period x 365 / assessment period) / total sewer length at the reference date x 100 *wOp24 = (wD28 x 365 / wH1) / wC1 x 100* *Note that " x 365 / assessment period" is a unit conversion expression and is not intended to be considered as extrapolation.* *This PI may be assessed for periods shorter than one year, but it is recommended that it be used only where data for the variables have been collected for at least a year. Where it has been used for shorter time periods, special consideration is required when used for comparisons, either internal or external to the undertaking. See section 7.1.*

SUBGROUP	*Code*	Indicator *(unit)*	Concept / *Processing rule* / *Comments*
	wOp25	Manhole chambers replacement, renewal, renovation or repair *(%/year)*	(Number of manhole chambers *replaced*, *renewed*, *renovated* or *repaired* during the assessment period x 365 / assessment period) / total number of manhole chambers at the reference date x 100 *wOp25 = (wD29 x 365 / wH1) / wC21 x 100* *Replacement of manhole covers only are not included.* *Note that " x 365 / assessment period" is a unit conversion expression and is not intended to be considered as extrapolation.* *This PI may be assessed for periods shorter than one year, but it is recommended that it be used only where data for the variables have been collected for at least a year. Where it has been used for shorter time periods, special consideration is required when used for comparisons, either internal or external to the undertaking. See section 7.1.*
	wOp26	Replacement of manhole covers *(%/year)*	(Number of manhole covers *replaced* during the assessment period x 365 / assessment period) / total number of manholes at the reference date x 100 *wOp26 = (wD30 x 365 / wH1) / wC21 x 100* *Note that " x 365 / assessment period" is a unit conversion expression and is not intended to be considered as extrapolation.* *This PI may be assessed for periods shorter than one year, but it is recommended that it be used only where data for the variables have been collected for at least a year. Where it has been used for shorter time periods, special consideration is required when used for comparisons, either internal or external to the undertaking. See section 7.1.*
	wOp27	Service connection *rehabilitation* *(%/year)*	(Number of *service connections replaced* or *renewed* during the assessment period x 365 / assessment period) / total number of service connections at the reference date x 100 *wOp27 = (wD31 x 365 / wH1) / wC29 x 100* *Note that " x 365 / assessment period" is a unit conversion expression and is not intended to be considered as extrapolation.* *This PI may be assessed for periods shorter than one year, but it is recommended that it be used only where data for the variables have been collected for at least a year. Where it has been used for shorter time periods, special consideration is required when used for comparisons, either internal or external to the undertaking. See section 7.1.*
PUMPS REHABILITATION	*wOp28*	Pump refurbishment *(%/year)*	(Total nominal power of pumps subject to *refurbishment* during the assessment period x 365 / assessment period) / total nominal power of pumps at the reference date x 100 *wOp28 = (wD32 x 365 / wH1) / wC8 x 100* *Refurbishment may include partial replacement.* *Note that " x 365 / assessment period" is a unit conversion expression and is not intended to be considered as extrapolation.* *This PI may be assessed for periods shorter than one year, but it is recommended that it be used only where data for the variables have been collected for at least a year. Where it has been used for shorter time periods, special consideration is required when used for comparisons, either internal or external to the undertaking. See section 7.1.*

SUBGROUP	*Code*	Indicator *(unit)*	Concept *Processing rule* *Comments*
	wOp29	Pump replacement *(%/year)*	(Total nominal power of pumps *replaced* during the assessment period x 365 / assessment period) / total nominal power of pumps at the reference date x 100 *wOp29 = (wD33 x 365 / wH1) / wC8 x 100* *This PI refers exclusively to full replacement of the pump and the engine.* *Note that " x 365 / assessment period" is a unit conversion expression and is not intended to be considered as extrapolation.* *This PI may be assessed for periods shorter than one year, but it is recommended that it be used only where data for the variables have been collected for at least a year. Where it has been used for shorter time periods, special consideration is required when used for comparisons, either internal or external to the undertaking. See section 7.1.*
INFLOW / INFILTRATION / EXFILTRATION (I/I/E)	*wOp30*	Inflow / Infiltration / Exfiltration (I/I/E) *(%)*	Volume of water entering sewers, from groundwater and *wrong connections* less the leakage from sewers into the ground / (*collected sewage* + *inflow* + *infiltration* – *exfiltration*) x 100, during the assessment period *wOp30=(wD35+wD36-wD37) /(wF1+wD35+wD36-wD37) x100* *This indicator may be assessed for periods shorter than one year, but special consideration is required when used for comparisons, either internal or external to the undertaking. See section 7.1.*
	wOp31	*inflow* *(m³/km/year)*	(Volume of water entering sewers from *wrong connections* during the assessment period x 365 / assessment period) / total sewer length at the reference date *wOp31 = (wD35 x 365 / wH1) / wC1* *Note that " x 365 / assessment period" is a unit conversion expression and is not intended to be considered as extrapolation.* *This indicator may be assessed for periods shorter than one year, but special consideration is required when used for comparisons, either internal or external to the undertaking. See section 7.1.*
	wOp32	*infiltration* *(m³/km/year)*	(Volume of water entering sewers from groundwater during the assessment period x 365 / assessment period) / total sewer length at the reference date *wOp32 = (wD36 x 365 / wH1) / wC1* *Note that " x 365 / assessment period" is a unit conversion expression and is not intended to be considered as extrapolation.* *This indicator may be assessed for periods shorter than one year, but special consideration is required when used for comparisons, either internal or external to the undertaking. See section 7.1.*
	wOp33	*exfiltration* *(m³/km/year)*	(Volume of leakage from sewers into the ground during the assessment period x 365 / assessment period) / total sewer length at the reference date *wOp33 = (wD37 x 365 / wH1) / wC1* *Note that " x 365 / assessment period" is a unit conversion expression and is not intended to be considered as extrapolation.* *This indicator may be assessed for periods shorter than one year, but special consideration is required when used for comparisons, either internal or external to the undertaking. See section 7.1.*

SUBGROUP	*Code*	Indicator *(unit)*	Concept *Processing rule* *Comments*
FAILURES	*wOp34*	Sewer blockages *(No./100 km sewer/year)*	(Number of *blockages* in sewers that occurred during the assessment period x 365 / assessment period) / total sewer length at the reference date x 100 *wOp34 = (wD38 x 365 / wH1) / wC1 x 100* *pumping station* *blockages shall not be included. Include blockages in* *service connections* *only where these are the responsibility of the wastewater undertaking.* *Note that " x 365 / assessment period" is a unit conversion expression and is not intended to be considered as extrapolation.* *This indicator may be assessed for periods shorter than one year, but special consideration is required when used for comparisons, either internal or external to the undertaking. See section 7.1.*
	wOp35	- sewer blockage locations *(No./100 km sewer/year)*	Number of individual locations in sewers where *blockages* occurred during the assessment period x 365 / assessment period) / total sewer length at the reference date x 100 *wOp35 = (wD39 x 365 / wH1) / wC1 x 100* *Locations where frequent blockages occurred should only be accounted once;* *pumping station* *blockages shall not be included. Include blockage locations in* *service connections* *only where these are the responsibility of the wastewater undertaking.* *Note that " x 365 / assessment period" is a unit conversion expression and is not intended to be considered as extrapolation.* *This indicator may be assessed for periods shorter than one year, but special consideration is required when used for comparisons, either internal or external to the undertaking. See section 7.1.*
	wOp36	Pumping station blockages *(No./pumping station/year)*	(Number of *blockages* that occurred in *pumping stations* during the assessment period x 365 / assessment period) / number of pumping stations at the reference date *wOp36 = (wD40 x 365 / wH1) / wC6 x 100* *Pumping station blockages include blockages in pumps and valves. Sewer blockages shall not be included.* *Note that " x 365 / assessment period" is a unit conversion expression and is not intended to be considered as extrapolation.* *This indicator may be assessed for periods shorter than one year, but special consideration is required when used for comparisons, either internal or external to the undertaking. See section 7.1.*
	wOp37	Flooding from sanitary sewers *(No./100 km sewer/year)*	(Number of *flooding* incidents related to *sanitary sewers* during the assessment period x 365 / assessment period) / total sewer length at the reference date x 100 *wOp37 = (wD41 x 365 / wH1) / wC1 x 100* *Include only incidents related to sanitary sewers that are the responsibility of the wastewater undertaking.* *Note that " x 365 / assessment period" is a unit conversion expression and is not intended to be considered as extrapolation.* *This indicator may be assessed for periods shorter than one year, but special consideration is required when used for comparisons, either internal or external to the undertaking. See section 7.1.*

SUBGROUP	Code	Indicator (unit)	Concept / Processing rule / Comments
	wOp38	Flooding from combined sewers *(No./100 km sewer/year)*	(Number of *flooding* incidents related to combined sewers during the assessment period x 365 / assessment period) / total sewer length at the reference date x 100 *wOp38 = (wD42 x 365 / wH1) / wC1 x 100* *Include only incidents related to combined sewers that are the responsibility of the wastewater undertaking.* *Note that " x 365 / assessment period" is a unit conversion expression and is not intended to be considered as extrapolation.* *This PI may be assessed for periods shorter than one year, but it is recommended that it be used only where data for the variables have been collected for at least a year. Where it has been used for shorter time periods, special consideration is required when used for comparisons, either internal or external to the undertaking. See section 7.1.*
	wOp39	Surface flooding *(No./100 km sewer/year)*	(Number of *surface flooding* incidents during the assessment period x 365 / assessment period) / total sewer length at the reference date x 100 *wOp39 = (wD43 x 365 / wH1) / wC1 x 100* *These include only surface flooding due to inadequacy of storm drainage system (combined sewers included) that is the responsibility of the wastewater undertaking. Inadequacy relates to all causes (e.g. design, operation, etc.).* *Note that " x 365 / assessment period" is a unit conversion expression and is not intended to be considered as extrapolation.* *This PI may be assessed for periods shorter than one year, but it is recommended that it be used only where data for the variables have been collected for at least a year. Where it has been used for shorter time periods, special consideration is required when used for comparisons, either internal or external to the undertaking. See section 7.1.*
	wOp40	Sewer collapses *(No./100 km sewer/year)*	(Number of sewer *collapses* during the assessment period x 365 / assessment period) / total sewer length at the reference date x 100 *wOp40 = (wD44 x 365 / wH1) / wC1 x 100* *Does not include collapses on service connections.* *Note that " x 365 / assessment period" is a unit conversion expression and is not intended to be considered as extrapolation.* *This indicator may be assessed for periods shorter than one year, but special consideration is required when used for comparisons, either internal or external to the undertaking. See section 7.1.*
	wOp41	Pump failures *(hours/pump/year)*	(Sum, for all pumps, of the number of hours when each pump was out of order during the assessment period x 365 / assessment period) / total number of pumps at the reference date *wOp41 = (wD34 x 365 / wH1) / wC7* *wastewater treatment plant pumps should not be included.* *Note that " x 365 / assessment period" is a unit conversion expression and is not intended to be considered as extrapolation.* *This indicator may be assessed for periods shorter than one year, but special consideration is required when used for comparisons, either internal or external to the undertaking. See section 7.1.*

SUBGROUP	Code	Indicator (unit)	Concept / Processing rule / Comments
	wOp42	Power failures *(hours/pumping station/year)*	(Sum, for all *pumping stations*, of the number of hours each pumping station was out of service due to power supply interruptions during the assessment period x 365 / assessment period) / total number of pumping stations at the reference date *wOp42 = (wD18 x 365 / wH1) / wC6* *This indicator assesses the time that pumping stations are out of service due to power failures, regardless of the power source used (normal or standby power generation system). This means that the time when a pumping station is reliant on standby power generation shall not be considered as a power failure. This indicator aims to measure the performance of the wastewater undertaking and not of the power supply undertaking.* *Note that " x 365 / assessment period" is a unit conversion expression and is not intended to be considered as extrapolation.* *This indicator may be assessed for periods shorter than one year, but special consideration is required when used for comparisons, either internal or external to the undertaking. See section 7.1.*
CSO CONTROL	*wOp43*	CSO control *(%)*	Number of *CSOs* controlled by dynamic or static *flow regulators* / total number of CSO devices x 100, at the reference date *wOp43 = wD45 / wC20 x 100*
WASTEWATER AND SLUDGE QUALITY MONITORING	*wOp44*	Wastewater quality tests carried out *(-/year)*	(Total number of tests carried out during the assessment period x 365 / assessment period) / total number of tests required by applicable standards or legislation during the assessment period *wOp44 = (wD46 x 365 / wH1) / wD57* *This indicator shall only account for tests that are the responsibility of the undertaking, carried out at points where quality monitoring is mandatory.* *Note that " x 365 / assessment period" is a unit conversion expression and is not intended to be considered as extrapolation.* *This indicator may be assessed for periods shorter than one year, but special consideration is required when used for comparisons, either internal or external to the undertaking. See section 7.1.*
	wOp45	- BOD tests *(-/year)*	(Number of BOD tests carried out during the assessment period x 365 / assessment period) / number of BOD tests required by applicable standards or legislation during the assessment period *wOp45 = (wD48 x 365 / wH1) / wD58* *This indicator shall only account for tests that are the responsibility of the undertaking, carried out at points where quality monitoring is mandatory.* *Note that " x 365 / assessment period" is a unit conversion expression and is not intended to be considered as extrapolation.* *This indicator may be assessed for periods shorter than one year, but special consideration is required when used for comparisons, either internal or external to the undertaking. See section 7.1.*

SUBGROUP	*Code*	Indicator (*unit*)	Concept / *Processing rule* / *Comments*
	wOp46	- COD tests (*-/year*)	(Number of COD tests carried out during the assessment period x 365 / assessment period) / number of COD tests required by applicable standards or legislation during the assessment period *wOp46 = (wD49 x 365 / wH1) / wD59* *This indicator shall only account for tests that are the responsibility of the undertaking, carried out at points where quality monitoring is mandatory.* *Note that " x 365 / assessment period" is a unit conversion expression and is not intended to be considered as extrapolation.* *This indicator may be assessed for periods shorter than one year, but special consideration is required when used for comparisons, either internal or external to the undertaking. See section 7.1.*
	wOp47	- TSS tests (*-/year*)	(Number of TSS tests carried out during the assessment period x 365 / assessment period) / number of TSS tests required by applicable standards or legislation during the assessment period *wOp47 = (wD50 x 365 / wH1) / wD60* *This indicator shall only account for tests that are the responsibility of the undertaking, carried out at points where quality monitoring is mandatory.* *Note that " x 365 / assessment period" is a unit conversion expression and is not intended to be considered as extrapolation.* *This indicator may be assessed for periods shorter than one year, but special consideration is required when used for comparisons, either internal or external to the undertaking. See section 7.1.*
	wOp48	- total phosphorus tests (*-/year*)	(Number of total phosphorus tests carried out during the assessment period x 365 / assessment period) / number of total phosphorus tests required by applicable standards or legislation during the assessment period *wOp48 = (wD51 x 365 / wH1) / wD61* *This indicator shall only account for tests that are the responsibility of the undertaking, carried out at points where quality monitoring is mandatory.* *Note that " x 365 / assessment period" is a unit conversion expression and is not intended to be considered as extrapolation.* *This indicator may be assessed for periods shorter than one year, but special consideration is required when used for comparisons, either internal or external to the undertaking. See section 7.1.*

SUBGROUP	*Code*	Indicator (unit)	Concept / *Processing rule* / *Comments*
	wOp49	- nitrogen tests *(-/year)*	(Number of nitrogen tests carried out during the assessment period x 365 / assessment period) / number of nitrogen tests required by applicable standards or legislation during the assessment period *wOp49 = (wD52 x 365 / wH1) / wD62* *Total number of tests assessing nitrogen and nitrogen compounds. This indicator shall only account for tests that are the responsibility of the undertaking, carried out at points where quality monitoring is mandatory.* *Note that " x 365 / assessment period" is a unit conversion expression and is not intended to be considered as extrapolation.* *This indicator may be assessed for periods shorter than one year, but special consideration is required when used for comparisons, either internal or external to the undertaking. See section 7.1.*
	wOp50	- faecal *E.coli* tests *(-/year)*	(Number of faecal *Escherichia coli* tests carried out during the assessment period x 365 / assessment period) / number of faecal *E.coli* tests required by applicable standards or legislation during the assessment period *wOp50 = (wD53 x 365 / wH1) / wD63* *This indicator shall only account for tests that are the responsibility of the undertaking, carried out at points where quality monitoring is mandatory.* *Note that " x 365 / assessment period" is a unit conversion expression and is not intended to be considered as extrapolation.* *This indicator may be assessed for periods shorter than one year, but special consideration is required when used for comparisons, either internal or external to the undertaking. See section 7.1.*
	wOp51	- other tests *(-/year)*	(Number of other tests carried out during the assessment period x 365 / assessment period) / number of other tests required by applicable standards or legislation during the assessment period *wOp51 = (wD54 x 365 / wH1) / wD64* *This indicator shall only account for tests that are the responsibility of the undertaking, carried out at points where quality monitoring is mandatory.* *Note that " x 365 / assessment period" is a unit conversion expression and is not intended to be considered as extrapolation.* *This indicator may be assessed for periods shorter than one year, but special consideration is required when used for comparisons, either internal or external to the undertaking. See section 7.1.*

SUBGROUP	Code	Indicator (unit)	Concept / Processing rule / Comments
	wOp52	Sludge tests carried out (-/year)	(Number of tests carried out to *sludge* produced during the assessment period x 365 / assessment period) / total number of tests required by applicable standards or legislation during the assessment period *wOp52 = (wD55 x 365 / wH1) / wD65* *This indicator shall only account for tests that are the responsibility of the undertaking, carried out at points where quality monitoring is mandatory.* *Note that " x 365 / assessment period" is a unit conversion expression and is not intended to be considered as extrapolation.* *This indicator may be assessed for periods shorter than one year, but special consideration is required when used for comparisons, either internal or external to the undertaking. See section 7.1.*
	wOp53	Industrial discharges tests carried out (-/year)	(Total number of tests carried out to industrial discharges in the sewer during the assessment period x 365 / assessment period) / total number of tests required by applicable standards or legislation during the assessment period *wOp53 = (wD56 x 365 / wH1) / wD66* *This indicator shall only account for tests carried out at points where quality monitoring is mandatory.* *Note that " x 365 / assessment period" is a unit conversion expression and is not intended to be considered as extrapolation.* *This indicator may be assessed for periods shorter than one year, but special consideration is required when used for comparisons, either internal or external to the undertaking. See section 7.1.*
VEHICLE AVAILABILITY	*wOp54*	Vehicle availability (No./100 km sewer)	Average daily number of vehicles available for *sewer system* related fieldwork / total sewer length x 100, at the reference date *wOp54 = wD67 / wC1 x 100* *In the case of multi-utilities, only the portion of vehicles related to wastewater service activities shall be considered.* *All vehicles that are available shall be accounted for, regardless of their ownership, undertaking or subcontractors. Vehicles used in the scope of outsourced activities shall be excluded. Heavy equipment shall not be included.*
SAFETY EQUIPMENT	*wOp55*	Gas detectors (No./employee)	Number of gas detectors (including: O_2, H_2S, CH_4 and CO_2) / operation and maintenance personnel, at the reference date *wOp55 = wD68 / wB12* *All gas detectors that are available shall be accounted for, regardless of their ownership, undertaking or subcontractors. Include both permanently installed and portable gas detectors that are calibrated.*
	wOp56	Permanently installed gas detectors (%)	Number of permanently installed gas detectors (including: O_2, H_2S, CH_4 and CO_2) / number of gas detectors x 100, at the reference date *wOp56 = wD69 / wD68 x 100*

7.6 Quality of service indicators (wQS)

Quality of service PIs measure the level of service provided to customers. Areas include level of service coverage, flooding and relations with customers, such as reply to requests, complaints, third party damage and traffic disruption caused by undertaking activities.

SUBGROUP	*Code*	Indicator (unit)	Concept / *Processing rule* / *Comments*
POPULATION SERVED	*wQS1*	Resident population connected to sewer system *(%)*	Resident population connected to the *sewer systems* managed by the undertaking / resident population x 100, at the reference date *wQS1 = wE4 / wE1 x 100*
	wQS2	Resident population served by WWTP *(%)*	Resident population served by *wastewater treatment plants* / resident population x 100, at the reference date *wQS2 = wE2 / wE1 x 100*
	wQS3	Resident population served by on-site systems *(%)*	Resident population served by *on-site systems* managed by the undertaking / resident population x 100, at the reference date *wQS3 = wE3 / wE1 x 100*
	wQS4	Resident population not served *(%)*	Resident population whose wastewater is neither collected nor treated / resident population x 100, at the reference date *wQS4 = wE7 / wE1 x 100*
TREATED WASTEWATER	*wQS5*	Treated wastewater in WWTP *(%)*	Volume of wastewater treated in *wastewater treatment plants* / *collected sewage* x 100, during the assessment period *wQS5 = (wA2 - wA11) / wF1 x 100* *This indicator may be assessed for periods shorter than one year, but special consideration is required when used for comparisons, either internal or external to the undertaking. See section 7.1.*
	wQS6	- preliminary treatment *(%)*	Volume of wastewater receiving only *preliminary treatment* at *wastewater treatment plants* / *collected sewage* x 100, during the assessment period *wQS6 = wA3 / wF1 x 100* *This indicator may be assessed for periods shorter than one year, but special consideration is required when used for comparisons, either internal or external to the undertaking. See section 7.1.*
	wQS7	- primary treatment *(%)*	Volume of wastewater receiving only *primary treatment* at *wastewater treatment plants* / *collected sewage* x 100, during the assessment period *wQS7 = wA5 / wF1 x 100* *This indicator may be assessed for periods shorter than one year, but special consideration is required when used for comparisons, either internal or external to the undertaking. See section 7.1.*

SUBGROUP	Code	Indicator (unit)	Concept / Processing rule / Comments
	wQS8	- secondary treatment (%)	Volume of wastewater receiving *secondary treatment* at *wastewater treatment plants* / *collected sewage* x 100, during the assessment period *wQS8 = wA7 / wF1 x 100* *This indicator may be assessed for periods shorter than one year, but special consideration is required when used for comparisons, either internal or external to the undertaking. See section 7.1.*
	wQS9	- tertiary treatment (%)	Volume of wastewater receiving *tertiary treatment* at *wastewater treatment plants* / *collected sewage* x 100, during the assessment period *wQS9 = wA9 / wF1 x 100* *This indicator may be assessed for periods shorter than one year, but special consideration is required when used for comparisons, either internal or external to the undertaking. See section 7.1.*
FLOODING	*wQS10*	Flooding of properties from sanitary sewers in dry weather (n.º/1000 properties/year)	(Number of properties affected by *flooding* from *sanitary sewers* in *dry weather*, during the assessment period x 365 / assessment period) / number of connected properties at the reference date x 1000 *wQS10 = (wF2 x 365 / wH1) / wC28 x 1000* *Only flooding from sanitary sewers that are the responsibility of the wastewater undertaking should be included. Flooding may affect properties that are not connected to the sewer system. These should be included.* *Note that " x 365 / assessment period" is a unit conversion expression and is not intended to be considered as extrapolation.* *This PI may be assessed for periods shorter than one year, but it is recommended that it be used only where data for the variables have been collected for at least a year. Where it has been used for shorter time periods, special consideration is required when used for comparisons, either internal or external to the undertaking. See section 7.1.*
	wQS11	Flooding of properties from sanitary sewers in wet weather (n.º/1000 properties/year)	(Number of properties affected by *flooding* from *sanitary sewers* in *wet weather*, during the assessment period x 365 / assessment period) / number of connected properties at the reference date x 1000 *wQS11 = (wF3 x 365 / wH1) / wC28 x 1000* *Only flooding from sanitary sewers that are the responsibility of the wastewater undertaking should be included. Flooding may affect properties that are not connected to the sewer system. These should be included.* *Note that " x 365 / assessment period" is a unit conversion expression and is not intended to be considered as extrapolation.* *This PI may be assessed for periods shorter than one year, but it is recommended that it be used only where data for the variables have been collected for at least a year. Where it has been used for shorter time periods, special consideration is required when used for comparisons, either internal or external to the undertaking. See section 7.1.*

SUBGROUP *Code*	Indicator *(unit)*	Concept *Processing rule* *Comments*
wQS12	Flooding of properties from combined sewers in dry weather *(n.º/1000 properties/year)*	(Number of properties affected by *flooding* from combined sewers in *dry weather*, during the assessment period x 365 / assessment period) / number of connected properties at the reference date x 1000 *wQS12 = (wF4 x 365 / wH1) / wC28 x 1000* *Only flooding from combined sewers that are the responsibility of the wastewater undertaking should be included. Flooding may affect properties that are not connected to the sewer system. These should be included.* *Note that " x 365 / assessment period" is a unit conversion expression and is not intended to be considered as extrapolation.* *This PI may be assessed for periods shorter than one year, but it is recommended that it be used only where data for the variables have been collected for at least a year. Where it has been used for shorter time periods, special consideration is required when used for comparisons, either internal or external to the undertaking. See section 7.1.*
wQS13	Flooding of properties from combined sewers in wet weather *(n.º/1000 properties/year)*	(Number of properties affected by *flooding* from combined sewers in *wet weather*, during the assessment period x 365 / assessment period) / number of connected properties at the reference date x 1000 *wQS13 = (wF5 x 365 / wH1) / wC28 x 1000* *Only flooding from combined sewers that are the responsibility of the wastewater undertaking should be included. Flooding may affect properties that are not connected to the sewer system. These should be included.* *Note that " x 365 / assessment period" is a unit conversion expression and is not intended to be considered as extrapolation.* *This PI may be assessed for periods shorter than one year, but it is recommended that it be used only where data for the variables have been collected for at least a year. Where it has been used for shorter time periods, special consideration is required when used for comparisons, either internal or external to the undertaking. See section 7.1.*
wQS14	Surface water flooding of properties in wet weather *(n.º/1000 properties/year)*	(Number of properties affected by *surface flooding* in *wet weather*, during the assessment period x 365 / assessment period) / number of connected properties at the reference date x 1000 *wQS14 = (wF6 x 365 / wH1) / wC28 x 1000* *Include only surface water flooding due to inadequacy of the storm drainage system (including combined sewers) that are the responsibility of the wastewater undertaking. (Inadequacy related to all causes).* *Note that " x 365 / assessment period" is a unit conversion expression and is not intended to be considered as extrapolation.* *This PI may be assessed for periods shorter than one year, but it is recommended that it be used only where data for the variables have been collected for at least a year. Where it has been used for shorter time periods, special consideration is required when used for comparisons, either internal or external to the undertaking. See section 7.1.*

SUBGROUP	Code	Indicator (unit)	Concept / Processing rule / Comments
INTERRUPTIONS	*wQS15*	Interruption of wastewater collection and transport services *(%)*	Sum, for the assessment period, of the number of properties affected by *service interruption* multiplied by the respective duration of interruptions in hours / (number of connected properties at the reference date x 24 x assessment period) x 100 *wQS15 = wF7 / (wC28 x 24 x wH1) x 100* *This indicator may be assessed for periods shorter than one year, but special consideration is required when used for comparisons, either internal or external to the undertaking. See section 7.1.*
REPLY TO CUSTOMER REQUESTS	*wQS16*	New service connection efficiency *(days/new connection)*	Total time spent for establishing new *service connections* / number of new service connections installed, during the assessment period *wQS16 = wF9 / wF8* *This indicator refers to new contracts, when new connections need to be established to sewers that already exist. The time is accounted for from the customer request up until the availability of the service.* *This indicator may be assessed for periods shorter than one year, but special consideration is required when used for comparisons, either internal or external to the undertaking. See section 7.1.*
	wQS17	Service connection repair time *(days/repaired connection)*	Total time spent for *repairing service connections* / total number of service connections repaired, during the assessment period *wQS17 = (wF11 x 365 / wH1) / wF10* *The time is accounted for since the time failures are reported until the service is resumed.* *This indicator may be assessed for periods shorter than one year, but special consideration is required when used for comparisons, either internal or external to the undertaking. See section 7.1.*
	wQS18	Average response time to empty septic tanks or pits *(days/request)*	Sum, for all requests, of the response time to empty *septic tanks* or contents of pit latrines / total number of requests for emptying septic tanks or contents of pit latrines, during the assessment period *wQS18 = wF24 / wF23* *Response time is the period in days from request to empty septic tanks or contents of pit latrines to the day the service is rendered.* *Note that " x 365 / assessment period" is a unit conversion expression and is not intended to be considered as extrapolation.* *This indicator may be assessed for periods shorter than one year, but special consideration is required when used for comparisons, either internal or external to the undertaking. See section 7.1.*
COMPLAINTS	*wQS19*	Total complaints *(No. /1000 inhab./year)*	(Total number of *complaints* related to wastewater system performance, during the assessment period x 365 / assessment period) / resident population at the reference date x 1000 *wQS19 = (wF12 x 365 / wH1) / wE1 x 1000* *Note that " x 365 / assessment period" is a unit conversion expression and is not intended to be considered as extrapolation.* *This indicator may be assessed for periods shorter than one year, but special consideration is required when used for comparisons, either internal or external to the undertaking. See section 7.1.*

SUBGROUP	Code	Indicator (unit)	Concept / Processing rule / Comments
	wQS20	- blockage complaints *(No. /1000 inhab./year)*	(Number of *complaints* as a result of *blockages*, during the assessment period x 365 / assessment period) / resident population at the reference date x 1000 *wQS20 = (wF13 x 365 / wH1) / wE1 x 1000* *Only complaints relating to the system that is the responsibility of the wastewater undertaking should be accounted for.* *Note that " x 365 / assessment period" is a unit conversion expression and is not intended to be considered as extrapolation.* *This indicator may be assessed for periods shorter than one year, but special consideration is required when used for comparisons, either internal or external to the undertaking. See section 7.1.*
	wQS21	- flooding complaints *(No. /1000 inhab./year)*	(Number of *complaints* as a result of *flooding*, during the assessment period x 365 / assessment period) / resident population at the reference date x 1000 *wQS21 = (wF14 x 365 / wH1) / wE1 x 1000* *Only complaints relating to the system that is the responsibility of the wastewater undertaking should be accounted for.* *Note that " x 365 / assessment period" is a unit conversion expression and is not intended to be considered as extrapolation.* *This indicator may be assessed for periods shorter than one year, but special consideration is required when used for comparisons, either internal or external to the undertaking. See section 7.1.*
	wQS22	- pollution incidents complaints *(No. /1000 inhab./year)*	(Number of *complaints* as a result of pollution incidents, during the assessment period x 365 / assessment period) / resident population at the reference date x 1000 *wQS22 = (wF15 x 365 / wH1) / wE1 x 1000* *Only complaints relating to the system that is the responsibility of the wastewater undertaking should be accounted for.* *Note that " x 365 / assessment period" is a unit conversion expression and is not intended to be considered as extrapolation.* *This indicator may be assessed for periods shorter than one year, but special consideration is required when used for comparisons, either internal or external to the undertaking. See section 7.1.*
	wQS23	- odour complaints *(No. /1000 inhab./year)*	(Number of *complaints* as a result of odours, during the assessment period x 365 / assessment period) / resident population at the reference date x 1000 *wQS23 = (wF16 x 365 / wH1) / wE1 x 1000* *Only complaints relating to the system that is the responsibility of the wastewater undertaking should be accounted for.* *Note that " x 365 / assessment period" is a unit conversion expression and is not intended to be considered as extrapolation.* *This indicator may be assessed for periods shorter than one year, but special consideration is required when used for comparisons, either internal or external to the undertaking. See section 7.1.*

SUBGROUP *Code*	Indicator *(unit)*	Concept *Processing rule* *Comments*
wQS24	- rodents related complaints *(No. /1000 inhab./year)*	(Number of *complaints* as a result of rodents, during the assessment period x 365 / assessment period) / resident population at the reference date x 1000 *wQS24 = (wF17 x 365 / wH1) / wE1 x 1000* *Only complaints relating to the system that is the responsibility of the wastewater undertaking should be accounted for.* *Note that " x 365 / assessment period" is a unit conversion expression and is not intended to be considered as extrapolation.* *This indicator may be assessed for periods shorter than one year, but special consideration is required when used for comparisons, either internal or external to the undertaking. See section 7.1.*
wQS25	- customer account related complaints *(No. /1000 inhab./year)*	(Number of *complaints* related to costumer accounts, during the assessment period x 365 / assessment period) / resident population at the reference date x 1000 *wQS25 = (wF18 x 365 / wH1) / wE1 x 1000* *Only complaints relating to the system that is the responsibility of the wastewater undertaking should be accounted for.* *Note that " x 365 / assessment period" is a unit conversion expression and is not intended to be considered as extrapolation.* *This indicator may be assessed for periods shorter than one year, but special consideration is required when used for comparisons, either internal or external to the undertaking. See section 7.1.*
wQS26	- other complaints *(No. /1000 inhab./year)*	(Number of other *complaints* during the assessment period x 365 / assessment period) / resident population at the reference date x 1000 *wQS26 = (wF19 x 365 / wH1) / wE1 x 1000* *Only complaints relating to the system that is the responsibility of the wastewater undertaking should be accounted for.* *Note that " x 365 / assessment period" is a unit conversion expression and is not intended to be considered as extrapolation.* *This indicator may be assessed for periods shorter than one year, but special consideration is required when used for comparisons, either internal or external to the undertaking. See section 7.1.*
wQS27	Response to complaints *(%)*	Number of responses given within the target time / total number of *complaints* relating to wastewater system performance x 100, during the assessment period *wQS27 = wF20 / wF12 x 100* *Applicable where there is a Guaranteed Standards Scheme (see section 8.3 - System profile).* *Note that " x 365 / assessment period" is a unit conversion expression and is not intended to be considered as extrapolation.* *This indicator may be assessed for periods shorter than one year, but special consideration is required when used for comparisons, either internal or external to the undertaking. See section 7.1.*

SUBGROUP	*Code*	Indicator (*unit*)	Concept / *Processing rule* / *Comments*
THIRD PARTY DAMAGES	*wQS28*	Responsibility for third party damages *(%)*	Number of occurrences related to accidents with persons not working for the undertaking or causing damage of third party properties, assets, vehicles or belongings / total number of accidents (that are the undertaking responsibility) x 100, during the assessment period *wQS28 = wF21 / wF22 x 100* *Note that " x 365 / assessment period" is a unit conversion expression and is not intended to be considered as extrapolation.* *This indicator may be assessed for periods shorter than one year, but special consideration is required when used for comparisons, either internal or external to the undertaking. See section 7.1.*
IMPACTS ON TRAFFIC	*wQS29*	Traffic disturbances *(km/traffic interruption)*	$\Sigma_{\text{for all traffic disturbances due to wastewater related activities}}$ (Traffic lane length unavailable x interruption duration in hours) / number of traffic disturbances due to wastewater related activities x assessment period x 24, during the assessment period *wQS29 = wF26 / wF25 x wH1 x 24* *Traffic disturbances include activities related to inspection, cleaning, maintenance, construction and rehabilitation works in the sewer system, where traffic is interrupted in one traffic lane at least. Roads where average traffic is below 5000 vehicles per day during the works shall not be considered.* *Note that " x 365 / assessment period" is a unit conversion expression and is not intended to be considered as extrapolation.* *This indicator may be assessed for periods shorter than one year, but special consideration is required when used for comparisons, either internal or external to the undertaking. See section 7.1.*

7.7 Economic and financial indicators (wFi)

Indicators in this group deal with the effectiveness and efficiency of the use of financial resources. Additionally, they provide a means to interpret the business management, indicating the company financial behaviour and ability to expand. The US dollar is used as a reference unit in order to allow easier international comparisons. Revenues, costs, composition of running costs per type of cost, per main function and per technical activity, composition of capital costs, investment and efficiency, leverage, liquidity and profitability indicators are included.

SUBGROUP	Code	Indicator (unit)	Concept / Processing rule / Comments
REVENUES	*wFi1*	Unit revenue *(US$/p.e./year)*	[(total operating revenues – capitalised costs of self-constructed assets) during the assessment period x 365 / assessment period] / total *population equivalent* served at the reference date *wFi1 = [(wG2 - wG33) x 365 / wH1] / wE8* *Note that wG2-wG33 = wG1 (total revenues). See section 4.6, Table 7 and Table 8 for definitions.* *Note that " x 365 / assessment period" is a unit conversion expression and is not intended to be considered as extrapolation.* *This PI may be assessed for periods shorter than one year, but it is recommended that it be used only where data for the variables have been collected for at least a year. Where it has been used for shorter time periods, special consideration is required when used for comparisons, either internal or external to the undertaking. See section 7.1.*
	wFi2	- service revenues *(%)*	Service revenues / total revenues x 100, during the assessment period *wFi2 = wG3 / wG1 x 100* *Service revenues include domestic, commercial and industrial service revenues.* *This PI may be assessed for periods shorter than one year, but it is recommended that it be used only where data for the variables have been collected for at least a year. Where it has been used for shorter time periods, special consideration is required when used for comparisons, either internal or external to the undertaking. See section 7.1.*
	wFi3	- other revenues *(%)*	Other revenues not coming from service / total revenues x 100, during the assessment period *wFi3 = (wG1 - wG3) / wG1 x 100* *This PI may be assessed for periods shorter than one year, but it is recommended that it be used only where data for the variables have been collected for at least a year. Where it has been used for shorter time periods, special consideration is required when used for comparisons, either internal or external to the undertaking. See section 7.1.*
	wFi4	Industrial revenues *(%)*	Industrial revenues / service revenues x 100, during the assessment period *wFi4 = wG4 / wG3 x 100* *This PI may be assessed for periods shorter than one year, but it is recommended that it be used only where data for the variables have been collected for at least a year. Where it has been used for shorter time periods, special consideration is required when used for comparisons, either internal or external to the undertaking. See section 7.1.*

SUBGROUP	*Code*	Indicator *(unit)*	Concept *Processing rule* *Comments*
COSTS	*wFi5*	Unit total cost per p.e.[7] *(US$/p.e./year)*	[(Running costs plus capital costs, related to wastewater treatment and *sewer system*, during the assessment period) x 365 / assessment period] / total *population equivalent* served by the wastewater service at the reference date *wFi5 = (wG5 x 365 / wH1) / wE8* *or wFi5 = [(wG6+wG7) x 365 / wH1] / wE8* *Note that " x 365 / assessment period" is a unit conversion expression and is not intended to be considered as extrapolation.* *This indicator may be assessed for periods shorter than one year, but special consideration is required when used for comparisons, either internal or external to the undertaking. See section 7.1.*
	wFi6	Unit total cost per length of sewer [7] *(US$/km sewer/year)*	[(Running costs plus capital costs, related to *sewer system*, during the assessment period) x 365 / assessment period] / total sewer length at the reference date *wFi6 = (wG5 x 365 / wH1) / wC1* *Note that " x 365 / assessment period" is a unit conversion expression and is not intended to be considered as extrapolation.* *This indicator may be assessed for periods shorter than one year, but special consideration is required when used for comparisons, either internal or external to the undertaking. See section 7.1.*
	wFi7	- unit running cost per p.e.[8] *(US$/p.e./year)*	(Running costs related to wastewater treatment and *sewer system* during the assessment period x 365 / assessment period) / total *population equivalent* served by the wastewater service at the reference date *wFi7 = (wG6 x 365 / wH1) / wE8* *Note that " x 365 / assessment period" is a unit conversion expression and is not intended to be considered as extrapolation.* *This indicator may be assessed for periods shorter than one year, but special consideration is required when used for comparisons, either internal or external to the undertaking. See section 7.1.*
	wFi8	- unit running cost per length of sewer [8] *(US$/km sewer/year)*	(Running costs related to *sewer system* during the assessment period x 365 / assessment period) / total sewer length at the reference date *wFi8 = (wG6 x 365 / wH1) / wC1* *Note that " x 365 / assessment period" is a unit conversion expression and is not intended to be considered as extrapolation.* *This indicator may be assessed for periods shorter than one year, but special consideration is required when used for comparisons, either internal or external to the undertaking. See section 7.1.*

7 Indicators *wFi5* and *wFi6* shall be used in alternative. *wFi6* applies if these costs are referred just to sewer systems.

8 Indicators *wFi7* and *wFi8* shall be used in alternative. *wFi8* applies if these costs are referred just to sewer systems.

SUBGROUP	Code	Indicator (unit)	Concept / Processing rule / Comments
	wFi9	- unit capital cost per p.e.[9]	Capital costs related to *sewer system* during the assessment period x 365 / assessment period) / total *population equivalent* served by the wastewater service at the reference date
	wFi10	- unit capital cost per length of sewer [9] *(US$/km sewer/year)*	(Capital costs related to wastewater treatment and *sewer system* during the assessment period x 365 / assessment period) / total *population equivalent* served by the wastewater service at the reference date *wFi10 = (wG7 x 365 / wH1) / wC1* *Note that " x 365 / assessment period" is a unit conversion expression and is not intended to be considered as extrapolation.* *This indicator may be assessed for periods shorter than one year, but special consideration is required when used for comparisons, either internal or external to the undertaking. See section 7.1.*
COMPOSITION OF RUNNING COSTS PER TYPE OF COSTS	*wFi11*	Internal manpower costs *(%)*	Internal manpower costs / running costs x 100, during the assessment period *wFi11 = wG9 / wG6 x 100* *This indicator may be assessed for periods shorter than one year, but special consideration is required when used for comparisons, either internal or external to the undertaking. See section 7.1.*
	wFi12	External services costs *(%)*	External services costs / running costs x 100, during the assessment period *wFi12 = wG10 / wG6 x 100* *This indicator may be assessed for periods shorter than one year, but special consideration is required when used for comparisons, either internal or external to the undertaking. See section 7.1.*
	wFi13	Energy costs *(%)*	Energy costs / running costs x 100, during the assessment period *wFi13 = wG11 / wG6 x 100* *This indicator may be assessed for periods shorter than one year, but special consideration is required when used for comparisons, either internal or external to the undertaking. See section 7.1.*
	wFi14	Materials, chemicals and other consumable costs *(%)*	Materials, chemicals and other consumable costs / running costs x 100, during the assessment period *wFi14 = wG12 / wG6 x 100* *This indicator may be assessed for periods shorter than one year, but special consideration is required when used for comparisons, either internal or external to the undertaking. See section 7.1.*

9 Indicators *wFi9* and *wFi10* shall be used in alternative. *wFi10* applies if these costs are referred just to sewer systems.

SUBGROUP	*Code*	Indicator *(unit)*	Concept *Processing rule* *Comments*
	wFi15	Other running costs *(%)*	(Leasing and rentals + taxes, levies & fees + exceptional earnings & losses + other operating expenditures) / running costs x 100, during the assessment period *wFi15 = (wG13+wG14+wG15+wG16) / wG6 x 100* *This indicator may be assessed for periods shorter than one year, but special consideration is required when used for comparisons, either internal or external to the undertaking. See section 7.1.*
COMPOSITION OF RUNNING COSTS PER MAIN FUNCTION (internal and outsourced)	*wFi16*	General management functions costs *(%)*	Running costs of general management functions / running costs x 100, during the assessment period *wFi16 = wG17 / wG6 x 100* *See Figure 5 and Table 1 to Table 6 for definition of functions. This indicator includes not only internal functions costs but also, when applicable, outsourced functions costs.* *This indicator may be assessed for periods shorter than one year, but special consideration is required when used for comparisons, either internal or external to the undertaking. See section 7.1.*
	wFi17	Human resources management functions costs *(%)*	Running costs of human resources management functions / running costs x 100, during the assessment period *wFi17 = wG18 / wG6 x 100* *See Figure 5 and Table 1 to Table 6 for definition of functions. This indicator includes not only internal functions costs but also, when applicable, outsourced functions costs.* *This indicator may be assessed for periods shorter than one year, but special consideration is required when used for comparisons, either internal or external to the undertaking. See section 7.1.*
	wFi18	Financial and commercial management functions costs *(%)*	Running costs of financial and commercial management functions / running costs x 100, during the assessment period *wFi18 = wG19 / wG6 x 100* *See Figure 5 and Table 1 to Table 6 for definition of functions. This indicator includes not only internal functions costs but also, when applicable, outsourced functions costs.* *This indicator may be assessed for periods shorter than one year, but special consideration is required when used for comparisons, either internal or external to the undertaking. See section 7.1.*
	wFi19	Customer service management costs *(%)*	Running costs of customer service management functions / running costs x 100, during the assessment period *wFi19 = wG20 / wG6 x 100* *See Figure 5 and Table 1 to Table 6 for definition of functions. This indicator includes not only internal functions costs but also, when applicable, outsourced functions costs.* *This indicator may be assessed for periods shorter than one year, but special consideration is required when used for comparisons, either internal or external to the undertaking. See section 7.1.*

SUBGROUP	Code	Indicator (unit)	Concept / *Processing rule* / *Comments*
	wFi20	Technical services functions costs *(%)*	Running costs of technical services functions / running costs x 100, during the assessment period *wFi20 = wG21 / wG6 x 100* *See Figure 5 and Table 1 to Table 6 for definition of functions. This indicator includes not only internal functions costs but also, when applicable, outsourced functions costs.* *This indicator may be assessed for periods shorter than one year, but special consideration is required when used for comparisons, either internal or external to the undertaking. See section 7.1.*
COMPOSITION OF RUNNING COSTS PER TECHNICAL FUNCTION ACTIVITY	*wFi21*	Wastewater treatment running costs *(%)*	Running costs of the wastewater treatment / running costs x 100, during the assessment period *wFi21 = wG22 / wG6 x 100* *This indicator may be assessed for periods shorter than one year, but special consideration is required when used for comparisons, either internal or external to the undertaking. See section 7.1.*
	wFi22	Sewer system running costs *(%)*	Running costs of *sewer systems* / running costs x 100, during the assessment period *wFi22 = wG23 / wG6 x 100* *This indicator may be assessed for periods shorter than one year, but special consideration is required when used for comparisons, either internal or external to the undertaking. See section 7.1.*
	wFi23	Wastewater quality monitoring running costs *(%)*	Running costs of wastewater quality monitoring / running costs x 100, during the assessment period *wFi23 = wG24 / wG6 x 100* *This indicator may be assessed for periods shorter than one year, but special consideration is required when used for comparisons, either internal or external to the undertaking. See section 7.1.*
	wFi24	Support services running costs *(%)*	Running costs of support services / running costs x 100, during the assessment period *wFi24 = wG25 / wG6 x 100* *This indicator may be assessed for periods shorter than one year, but special consideration is required when used for comparisons, either internal or external to the undertaking. See section 7.1.*
COMPOSITION OF CAPITAL COSTS	*wFi25*	Depreciation costs *(%)*	Depreciation costs / capital costs x 100, during the assessment period *wFi25 = wG26 / wG7 x 100* *This indicator may be assessed for periods shorter than one year, but special consideration is required when used for comparisons, either internal or external to the undertaking. See section 7.1.*
	wFi26	Net interest costs *(%)*	(interest expenses costs – interest income) / capital costs x 100, during the assessment period *wFi26 = wG29 / wG7 x 100* *or wFi26 = (wG27 - wG28) / wG7 x 100* *This indicator may be assessed for periods shorter than one year, but special consideration is required when used for comparisons, either internal or external to the undertaking. See section 7.1.*

SUBGROUP	Code	Indicator (unit)	Concept / Processing rule / Comments
INVESTMENT	*wFi27*	Unit investment *(US$/p.e./year)*	(Cost of investments (expenditures for sewers, treatment plants and equipments) during the assessment period x 365 / assessment period) / *population equivalent* served by the wastewater service at the reference date *wFi27 = (wG30 x 365 / wH1) / wE8* *Note that " x 365 / assessment period" is a unit conversion expression and is not intended to be considered as extrapolation.* *This PI may be assessed for periods shorter than one year but the resulting values can be misleading. Thus it is recommended to be assessed using annual data for periods of more than one year. Where it has been used for shorter time periods, special attention shall be paid when this PI is used for comparisons, either internal or external to the undertaking. See section 7.1.*
	wFi28	- investments for new assets and reinforcement of existing assets *(%)*	Cost of investments for new assets and *reinforcement* of existing assets / total cost of the investments x 100, during the assessment period *wFi28 = wG31 / wG30 x 100* *Only added capacity investment shall be accounted for.* *This PI may be assessed for periods shorter than one year but the resulting values can be misleading. Thus it is recommended to be assessed using annual data for periods of more than one year. Where it has been used for shorter time periods, special attention shall be paid when this PI is used for comparisons, either internal or external to the undertaking. See section 7.1.*
	wFi29	- investments for asset replacement and renovation *(%)*	Cost of investments for the *replacement* and *renovation* ("like for like") of existing assets / total cost of the investments x 100, during the assessment period *wFi29 = wG32 / wG30 x 100* *"like for like" means that it will provide the same functionality. This PI may be assessed for periods shorter than one year but the resulting values can be misleading. Thus it is recommended to be assessed using annual data for periods of more than one year. Where it has been used for shorter time periods, special attention shall be paid when this PI is used for comparisons, either internal or external to the undertaking. See section 7.1.*
EFFICIENCY INDICATORS	*wFi30*	Total cost coverage ratio *(-)*	Total revenues / total costs, during the assessment period *wFi30 = wG1 / wG5* *This indicator may be assessed for periods shorter than one year, but special consideration is required when used for comparisons, either internal or external to the undertaking. See section 7.1.*
	wFi31	Operating cost coverage ratio *(-)*	Total revenues / running costs, during the assessment period *wFi31 = wG1 / wG6* *This indicator may be assessed for periods shorter than one year, but special consideration is required when used for comparisons, either internal or external to the undertaking. See section 7.1.*

SUBGROUP	Code	Indicator (unit)	Concept / Processing rule / Comments
	wFi32	Delay in accounts receivable *(days equivalent)*	(Accounts receivable from wastewater at the reference date / service revenues during the assessment period) x assessment period *wFi32 = wG34 / wG3 x wH1* *Notice that "x wH1" is a unit conversion expression.* *This PI may be assessed for periods shorter than one year, but it is recommended that it be used only where data for the variables have been collected for at least a year. Where it has been used for shorter time periods, special consideration is required when used for comparisons, either internal or external to the undertaking. See section 7.1.*
	wFi33	Investment ratio *(-)*	Investments subject to depreciation / depreciation costs, during the assessment period *wFi33 = wG35 / wG26* *This PI may be assessed for periods shorter than one year, but it is recommended that it be used only where data for the variables have been collected for at least a year. Where it has been used for shorter time periods, special consideration is required when used for comparisons, either internal or external to the undertaking. See section 7.1.*
	wFi34	Contribution of internal sources to investment = CTI *(%)*	Investments financed by the cash-flow / total investments x 100, during the assessment period *wFi34 = wG36 / wG30 x 100* *This PI may be assessed for periods shorter than one year, but it is recommended that it be used only where data for the variables have been collected for at least a year. Where it has been used for shorter time periods, special consideration is required when used for comparisons, either internal or external to the undertaking. See section 7.1.*
	wFi35	Average age of tangible assets *(%/year)*	Depreciated historical value of tangible assets / historical value of tangible assets x 100, during the year *wFi35 = wG37 / wG38 x 100* *This PI must only be assessed using annual data.*
	wFi36	Average depreciation ratio *(-/year)*	Depreciation costs / historical value of tangible assets, during the year *wFi36 = wG26 / wG38* *This PI must only be assessed using annual data.*
	wFi37	Late payments ratio *(-/year)*	[1 - (annual debt from customers / annual amount billed)], during the year *wFi37 = [1 - (wG39 / wG40)]* *This PI must only be assessed using annual data.*
	wFi38	Inventory value *(-/year)*	Value of overall inventory at the end of the fiscal year / total operating revenues during the year *wFi38 = wG47 / wG2* *This PI must only be assessed using annual data.*

SUBGROUP	*Code*	Indicator *(unit)*	Concept / *Processing rule* / *Comments*
LEVERAGE INDICATORS	*wFi39*	Debt service coverage ratio = DSC *(%)*	Cash-flow / financial debt service x 100, during the assessment period *wFi39 = wG41 / wG42 x 100* *This PI may be assessed for periods shorter than one year, but it is recommended that it be used only where data for the variables have been collected for at least a year. Where it has been used for shorter time periods, special consideration is required when used for comparisons, either internal or external to the undertaking. See section 7.1.*
	wFi40	Debt equity ratio *(-/year)*	Total debt / *shareholders'* equity, at the end of the fiscal year *wFi40 = wG43 / wG44* *This PI must only be assessed using annual data. Undertakings with high debt service coverage ratio tend to have low debt equity ratio. It is also affected by revaluation of fixed assets and therefore caution should be exercised when using this ratio.*
LIQUIDITY INDICATOR	*wFi41*	Current ratio *(-)*	Current assets / current liabilities, at the reference date *wFi41 = wG45 / wG49* *This PI may be assessed for periods shorter than one year, but it is recommended that it be used only where data for the variables have been collected for at least a year. Where it has been used for shorter time periods, special consideration is required when used for comparisons, either internal or external to the undertaking. See section 7.1.*
PROFITABILITY INDICATORS	*wFi42*	Return on net fixed assets *(%/year)*	Operating income / (historical value of tangible assets minus depreciated historical value of tangible assets) x 100, during the year *wFi42 =wG50 / (wG38 - wG37) x 100* *This PI must only be assessed using annual data.*
	wFi43	Return on equity *(%)*	Net income (after interest payment and taxes) / *shareholders'* equity x 100, during the year *wFi43 = wG52 / wG44 x 100* *This PI must only be assessed using annual data.*
	wFi44	Return on capital employed *(%)*	(Operating income (EBIT) - related taxes) / total assets x 100, during the year *wFi44 = (wG50 - wG51) / wG46 x 100* *This PI must only be assessed using annual data.*
	wFi45	Asset turnover ratio *(-/year)*	Service revenues / total assets, during the year *wFi45 = wG3 / wG46* *This PI must only be assessed using annual data.*

8. CONTEXT INFORMATION

8.1 Introduction

Wastewater undertakings act within a variety of contexts related to: the characteristics of the region in which they operate; the characteristics of the wastewater infrastructure systems they manage; and also within the framework and rules that govern the undertaking organisation, financing and management. The context also includes the constraints, legislative and regional, and other Institutional constraints that define the local structure of the wastewater service industry. Within the latter in particular, constraints on freedom of action often control the behaviour of the undertaking and restrict actions. Clearly these contextual factors impact significantly on the performance of the undertaking. Thus any assessment of performance in terms of service delivered needs to take into account the context in which the undertaking has to operate. This includes the most relevant characteristics of the infrastructure assets, resource systems and the nature of the region.

In this manual the PIs are complemented by Context information (CI) as outlined in Section 4.7 and shown in Figure 6. The CI data are organised into sectors for: the undertaking profile, the system profile and the region profile. The data items used in the CI lists have been devised to collectively highlight important aspects of the three sectors that may be significant for the performance of an undertaking. Some of the individual data comprising the set of context information may not be controlled by the undertaking but still affect its' performance in terms of service delivery. The CI data may be useful to identify wastewater undertakings that have similar contextual characteristics and can therefore be compared. The CI data can also help to explain differences between performances when comparing undertakings (WERF, 1997).

The undertaking profile outlines the framework of the organisation. The system profile focuses mainly on the type of wastewater service provided, on the physical assets, on the technological means used and on the customers. The system profile requires more detail than the other profiles because it also contains descriptive information that is helpful for the interpretation of the PIs. The region profile provides information about the demographic, economic, geographical and environmental context.

In common with the PI variables, some of the CI data may be obtained to cover various time periods. This requires CI data processing rules that convert the base data into equivalent annual information. As an example, *sludge disposed* is defined as a CI variable and can be assessed using an assessment period other than the year, although it is expressed in units of ton DS/year. In contrast, the CI for *Sludge treated in WWTP* is not defined as a variable, and it is also expressed in units of ton DS/year. Data for this must therefore be obtained annually.

8.2 Undertaking profile

CONTEXT INFORMATION	Undertaking profile data	*(unit/option)*	Concept *Comments*	*variable*
	Undertaking identification	*(-)*	Name of the wastewater undertaking, contact person, address, telephone, fax, e-mail	
	Geographical scope		Scope of activity of the organisation as a whole	
	- country	*(yes/no)*	*One single choice is to be selected. Where "yes", specify designation of the area covered by the service.*	
	- state	*(yes/no)*		
	- region	*(yes/no)*		
	- local	*(yes/no)*		

CONTEXT INFORMATION	Undertaking profile data *(unit/option)*	Concept *Comments* *variable*
	Type of activity - Wastewater collection *(yes/no)* - Wastewater transport *(yes/no)* - Wastewater treatment *(yes/no)* - Wastewater disposal *(yes/no)* - stormwater collection *(yes/no)* - Stormwater transport *(yes/no)* - Stormwater treatment *(yes/no)* - Stormwater disposal *(yes/no)* and: - water supply *(yes/no)* - electricity *(yes/no)* - solid waste *(yes/no)* - gas *(yes/no)* - district heating *(yes/no)* - other (specify): *(yes/no)*	Scope of activity of the organisation as a whole *Multiple choices are valid.*
	Type of assets ownership - public *(yes/no)* - private *(yes/no)* - public-private *(yes/no)*	Ownership of the wastewater infrastructure *One single choice is to be selected.*
	Type of operations - public *(yes/no)* - private *(yes/no)* - public-private partnership *(yes/no)*	Type of operational management of the undertaking *One single choice is to be selected.*
	Total personnel *(No.)*	Total number of full time equivalent employees of the wastewater undertaking *wB1*
	Annual working time per employee *(hours/employee/year)*	Number of annual working hours of all the undertaking employees (employed hours gross minus absence for holidays) / total personnel *wB26 / wB1*
	Annual revenues *(US$/year)*	Operating annual revenues minus capitalised costs of self-constructed assets, regarding the wastewater service *(wG1)* *See section 4.6 - Financial aspects, Table 7 and Table 8. Equivalent to wG1 for one year period.*
	Annual costs *(US$/year)*	Annual costs, including capital and running costs, regarding the wastewater service *(wG5)* *See section 4.6 - Financial aspects, Table 7 and Table 8. Equivalent to wG5 for one year period.*

CONTEXT INFORMATION	Undertaking profile data *(unit/option)*	Concept *Comments* *variable*
	Outsourcing costs *(US$/year)*	Annual costs of outsourcing, regarding the wastewater service
	- General management outsourcing costs *(US$/year)*	Annual costs of the general management function that is outsourced, regarding the wastewater service *See section 4.5 - Organisational functions and definitions, Table 1.*
	- Human resources outsourcing costs *(US$/year)*	Annual cost of the human resources management function that is outsourced, regarding the wastewater service *See section 4.5 - Organisational functions and definitions, Table 2.*
	- Financial and commercial outsourcing costs *(US$/year)*	Annual cost of the financial and commercial management function that is outsourced, regarding the wastewater service *See section 4.5 -Organisational functions and definitions, Table 3.*
	- Costumer service outsourcing costs *(US$/year)*	Annual cost of the costumer service function that is outsourced, regarding the wastewater service *See section 4.5 - Organisational functions and definitions, Table 4.*
	- Planning and design outsourcing costs *(US$/year)*	Annual cost of the planning and design function that is outsourced (e.g. use of consultants), regarding the wastewater service *See section 4.5 - Organisational functions and definitions, Table 5.*
	- Construction outsourcing costs *(US$/year)*	Annual cost of the construction function that is outsourced (proportion by value of new works and asset *refurbishment* undertaken by external contractor, including new *service connections* and extensions of network), regarding the wastewater service *See section 4.5 - Organisational functions and definitions, Table 5.*
	- Operation and maintenance outsourcing costs *(US$/year)*	Annual cost of the operation and maintenance function that is outsourced, regarding the wastewater service *See section 4.5 - Organisational functions and definitions, Table 6.*
	- Laboratory services outsourcing costs *(US$/year)*	Annual cost of the laboratory services function that is outsourced, regarding the wastewater service *See section 4.5 - Organisational functions and definitions, Table 6.*
	Average annual investment *(US$/year)*	Average annual cost of the investments (expenditure for infrastructure and associated equipment), regarding the wastewater service *(wG30)* *Equivalent to wG30 for one year period.*
	Income taxes *(US$/year)*	All taxes and levies on gross income related to *sewer system* and wastewater treatment activities, regarding the wastewater service *See section 4.6 - Financial aspects, Table 8.*

8.3 System profile

CONTEXT INFORMATION	System profile data *(unit/option)*	Concept *Comments*	*variable*
SERVICE DATA	Types of systems managed by the undertaking	*Multiple choices are valid.*	
	- sewers (excluding *interceptor sewers*) *(yes/no)*		
	- *interceptor sewers* *(yes/no)*		
	- *wastewater treatment plants* *(yes/no)*		
	- *on-site systems* *(yes/no)*		
	- *stormwater* only facilities *(yes/no)*		
	Resident population *(Inhab.)*	Total population living permanently in the area that is the responsibility of the wastewater undertaking	*wE1*
	- resident population connected to sewer systems *(Inhab.)*	Resident population connected to *sewer systems* managed by the undertaking	*wE4*
	- resident population connected to interceptor sewers *(Inhab.)*	Resident population connected to *interceptor sewers* managed by the undertaking	
	- resident population served by wastewater treatment plants *(Inhab.)*	Resident population served by *wastewater treatment plants* managed by the undertaking	*wE2*
	- resident population served by on-site systems *(Inhab.)*	Resident population served by *on-site systems* managed by the undertaking (e.g. *septic tanks*, latrines)	*wE3*
	peak population *(Inhab.)*	Maximum number of resident and non-resident population in the area that is the responsibility of the wastewater undertaking	
	Peak pollution load in WWTP *(p.e.)*	Maximum sum of daily BOD_5 peak load entering *wastewater treatment plants*, based on the *population equivalent* definition	
	Daily collected sewage *(m^3/day)*	Average daily *collected sewage*, corresponding to the volume of domestic, commercial and industrial inputs to the *sewer system*	*wF1 / wH1*
	Industrial wastewater *(%)*	Volume of collected *industrial wastewater* / *collected sewage* x 100	
	Imported wastewater *(m^3/day)*	Annual volume of *imported wastewater* during the year / 365	
	Exported wastewater *(m^3/day)*	Annual volume of *exported wastewater* during the year / 365	

CONTEXT INFORMATION	System profile data *(unit/option)*	Concept *Comments*	*variable*
	Level of wastewater treatment		
	- preliminary treated wastewater *(m^3/day)*	Annual average daily volume of wastewater receiving only *preliminary treatment*	*wA3 / wH1*
	- primary treated wastewater *(m^3/day)*	Annual average daily volume of wastewater receiving only *primary treatment*	*wA5 / wH1*
	- secondary treated wastewater *(m^3/day)*	Annual average daily volume of wastewater receiving *secondary treatment*	*wA7 / wH1*
	- tertiary treated wastewater *(m^3/day)*	Annual average daily volume of wastewater receiving *tertiary treatment*	*wA9 / wH1*
	- wastewater without treatment *(m^3/day)*	Annual average daily volume of wastewater discharged without treatment (does not include *overflow* discharge volumes)	
	Sludge produced in WWTP *(ton DS/year)*	Dry weight of *sludge* produced in *wastewater treatment plants* managed by the undertaking *Equivalent to wA13 for one year period.*	*(wA13)*
	Sludge treated in WWTP *(ton DS/year)*	Dry weight of *sludge* (from all sources) treated in *wastewater treatment plants* managed by the undertaking	
	Sludge disposed *(ton DS/year)*	Dry weight of *sludge* handled by the undertaking that is disposed in landfill, incinerated or by any other way of disposal *Equivalent to wA16 for one year period.*	*(wA16)*
	Screenings and grit produced *(ton/year)*	Drained weight of screenings and grit removed in *wastewater treatment plants* and in *pumping stations* *Equivalent to wA22 for one year period.*	*(wA22)*
CUSTOMER SERVICE	Existence of system to record complaints *(yes/no)*	Existence of registers that record total number of verbal and written *complaints*, enabling nature of complaint to be determined by scrutinising individual entries	
	Existence of formalised system to record complaints for service quality monitoring and asset management purposes *(yes/no)*	Existence of *complaint* recording and data processing system that is used for resolving complaints, monitoring of service quality and performance and asset management planning	
	Existence of a guaranteed standards scheme *(yes/no)*	Existence of a guaranteed standards scheme (GSS) that establishes the rights of the customers, including at least: maximum time to get a new connection and to *repair* an existing one; maximum time of written responses to *complaints*; appointment times to attend customers	

CONTEXT INFORMATION	System profile data	(unit/option)	Concept *Comments*	variable
CATCHMENT AREA	Catchment area	*(km^2)*	Extent of area receiving the waters feeding a part or the totality of a *drain runoff* or channel/sewer network	
	Impermeable catchment area	*(km^2)*	Area within the catchment area that is impermeable	
PHYSICAL ASSETS - SEWER NETWORK	Wastewater systems	*(No.)*	Number of autonomous wastewater systems managed by the undertaking at the reference date	
	Total sewer length	*(km)*	Total length of sewers managed by the undertaking at the reference date *service connections* excluded.	*wC1*
	- combined sewers length	*(km)*	Length of combined sewers managed by the undertaking at the reference date	
	- sanitary sewers length	*(km)*	Length of *sanitary sewers* managed by the undertaking at the reference date	
	- stormwater sewers length	*(km)*	Length of *stormwater* sewers managed by the undertaking at the reference date	
	- pump mains length	*(km)*	Length of *pump mains* managed by the undertaking at the reference date	
	- other sewers length	*(km)*	Length of other sewers (e.g. *small-bore sewers*, *settled sewers*, etc.) managed by the undertaking at the reference date	
	Expansion of sewer network	*(km/year)*	Annual increase in sewer length that is the responsibility of the wastewater undertaking (*service connections* excluded)	
	Sewer materials			
	- clay sewers	*(%)*	Length of clay sewers / total sewer length x 100	
	- asbestos cement sewers	*(%)*	Length of asbestos sewers / total sewer length x 100	
	- concrete sewers	*(%)*	Length of concrete sewers / total sewer length x 100	
	- polyvinyl chlorine sewers	*(%)*	Length of polyvinyl chlorine (PVC) sewers / total sewer length x 100	
	- polyethylene sewers	*(%)*	Length of polyethylene sewers / total sewer length x 100	
	- iron sewers	*(%)*	Length of iron sewers / total sewer length x 100	
	- steel sewers	*(%)*	Length of steel sewers / total sewer length x 100	
	- stone sewers	*(%)*	Length of stone sewers / total sewer length x 100	
	- brick sewers	*(%)*	Length of brick sewers / total sewer length x 100	
	- other known material sewers	*(%)*	Length of other known material sewers / total sewer length x 100	

CONTEXT INFORMATION	System profile data	(unit/option)	Concept *Comments*	*variable*
	- unknown material sewers	(%)	Length of unknown material sewers / total sewer length x 100	
	Sewer diameters or equivalent			
	Dia ≤ 150 mm	(%)	Length of sewers with diameter ≤ 150mm / total sewer length x 100	
	150 < Dia ≤ 450 mm	(%)	Length of sewers with 150 mm < diameter ≤ 450 mm / total sewer length x 100	
	450 < Dia ≤ 900 mm	(%)	Length of sewers with 450 mm < diameter ≤ 900 mm / total sewer length x 100	
	900 < Dia ≤ 1200 mm	(%)	Length of sewers with 900 mm < diameter ≤ 1200 mm / total sewer length x 100	
	1200 < Dia ≤ 2200 mm	(%)	Length of sewers with 1200 mm < diameter ≤ 2200 mm / total sewer length x 100	
	Dia > 2200 mm	(%)	Length of sewers with diameter > 2200 mm / total sewer length x 100	
	Unknown diameter sewers	(%)	Length of sewers with unknown diameters / total sewer length x 100	
	Sewer age		*In the items below, total sewer length is referred to the reference date.*	
	- sewers laid after 1995	(%)	Length of sewers laid after 1995 inclusive / total sewer length x 100	
	- sewers laid between 1985 and 1994 inclusive	(%)	Length of sewers laid between 1985 and 1994 inclusive / total sewer length x 100	
	- sewers laid between 1975 and 1984 inclusive	(%)	Length of sewers between 1975 and 1984 inclusive / total sewer length x 100	
	- sewers laid between 1950 and 1974 inclusive	(%)	Length of sewers between 1950 and 1974 inclusive / total sewer length x 100	
	- sewers laid between 1925 and 1949 inclusive	(%)	Length of sewers laid between 1925 and 1949 inclusive / total sewer length x 100	
	- sewers laid before 1925	(%)	Length of sewers laid before 1925 / total sewer length x 100	
	- unknown age sewers	(%)	Length of unknown age sewers / total sewer length x 100	
	Average age of sewer system	(years)	Average age of the *sewer system* based on the distribution of ages and lengths *This information is an rough value calculated from the above, ignoring the category of unknown age.*	

CONTEXT INFORMATION	System profile data (unit/option)		Concept *Comments*	variable
	Manholes			
	- manholes	*(No.)*	Number of manhole chambers in the *sewer system*	*wC21*
	- manholes maximum depth	*(m)*	Maximum manhole chamber depth in the *sewer system*	
	- manholes median depth	*(m)*	Median manhole chamber depth in the *sewer system*	
	- manholes minimum depth	*(m)*	Minimum manhole chamber depth in the *sewer system*	
	Gully pots	*(No.)*	Number of *gully pots* in the *sewer system*	wC22
	Sewer overflows	*(No.)*	Number of sewer *overflows*	
	- combined sewer overflow devices	*(No.)*	Number of *combined sewer overflow* (*CSO*) devices in the *sewer system*	*wC20*
	- storm overflow devices	*(No.)*	Number of *storm overflow devices* in the *sewer system*	
	Outfalls	*(No.)*	Number of outfalls managed by the undertaking into non-*tidal waters*	
	Marine outfalls	*(No.)*	Number of marine outfalls managed by the undertaking into *tidal waters*	
	Tidal sewers	*(km)*	Length of sewers subject to tidal influence	
SERVICE CONNECTIONS	Service connections	*(No.)*	Total number of *service connections*	wC29
	- domestic connections	*(No.)*	Number of *service connections* from residential properties/buildings *Exclude on-site system service connections.*	
	- industrial connections	*(No.)*	Number of *service connections* from industrial facilities without *industrial pre-treatment*	
	- on-site treatment connections	*(No.)*	Number of on-site (domestic, commercial and industrial) *service connections* with some *pre-treatment*	
	- other connections	*(No.)*	Number of *service connections* from non-residential properties/buildings (i.e.: commercial buildings, hospitals, schools), excluding industrial and on-site treatment service connections	
	Service connections density	*(No./km)*	Number of *service connections* / total sewer length	wC29 / wC1
WASTEWATER STORAGE	Storage tanks	*(No.)*	Number of *storage tanks* (in or off line) in the *sewer system*	*wC23*
	Total storage volume	*(m^3)*	Total volume of *storage tanks* (in or off line) in the *sewer system*	

CONTEXT INFORMATION	System profile data *(unit/option)*	Concept *Comments* *variable*
STORMWATER STORAGE	Stormwater storage tanks *(No.)*	Number of *stormwater* only *storage tanks* (in or off line) in the *sewer system*
	Other stormwater storage units *(No.)*	Number of other *stormwater* only *storage* units managed by the undertaking (e.g. surface ponds)
PUMPING STATIONS	Wastewater system pumping stations *(No.)*	Number of wastewater *pumping stations* in the *sewer system*, including *wastewater treatment plant* inlet pumping stations *wC6*
	Wastewater system pumps power *(kW)*	Total nominal power of the wastewater pumps in the *sewer system*, including *wastewater treatment plants* inlet *pumping stations* *wC8*
	Pumped wastewater *(%)*	Annual volume of pumped wastewater / *collected sewage* x 100
	Power supply interruptions *(hours)*	Sum, for all *pumping stations*, of the number of hours each pumping station is out of service due to power supply interruptions or is reliant on standby power generation, during the assessment period
WASTEWATER TREATMENT PLANTS	Total number of WWTP *(No.)*	Total number of *wastewater treatment plants* managed by the undertaking
	- Total separate system WWTP *(No.)*	Number of *wastewater treatment plants* managed by the undertaking, receiving wastewater from *separate sewer systems*
	- Total combined system WWTP *(No.)*	Number of *wastewater treatment plants* managed by the undertaking, receiving wastewater from *combined sewer systems*
	- Total mixed system (separate+combined) WWTP *(No.)*	Number of *wastewater treatment plants* managed by the undertaking, receiving wastewater from mixed (separate+combined) *sewer systems*
	Number of WWTP ≤ 500 p.e. *(No.)*	Number of *wastewater treatment plants* managed by the undertaking treating up to 500 p.e. daily
	Number of 500 < WWTP ≤ 2000 p.e. *(No.)*	Number of *wastewater treatment plants* managed by the undertaking treating more than 500 p.e. and up to 2000 p.e. daily
	Number of 2000 < WWTP ≤ 5000 p.e. *(No.)*	Number of *wastewater treatment plants* managed by the undertaking treating more than 2000 p.e. and up to 5000 p.e. daily
	Number of 5000 < WWTP ≤ 10000 p.e. *(No.)*	Number of *wastewater treatment plants* managed by the undertaking treating more than 5000 p.e. and up to 10000 p.e. daily
	Number of 10000 < WWTP ≤ 25000 p.e. *(No.)*	Number of *wastewater treatment plants* managed by the undertaking treating more than 10000 p.e. and up to 25000 p.e. daily
	Number of 25000 < WWTP ≤ 50000 p.e. *(No.)*	Number of *wastewater treatment plants* managed by the undertaking treating more than 25000 p.e. and up to 50000 p.e. daily

CONTEXT INFORMATION	System profile data (unit/option)		Concept *Comments* variable
	Number of 50000 < WWTP ≤ 100000 p.e.	(No.)	Number of *wastewater treatment plants* managed by the undertaking treating more than 50000 p.e. and up to 100000 p.e. daily
	Number of 100000 < WWTP ≤ 150000 p.e.	(No.)	Number of *wastewater treatment plants* managed by the undertaking treating more than 100000 p.e. and up to 150000 p.e. daily
	Number of WWTP > 150000 p.e.	(No.)	Number of *wastewater treatment plants* managed by the undertaking treating more than 150000 p.e. daily
	Treatment capacity		
	- preliminary treatment capacity	p.e.	Installed *preliminary treatment* capacity *For this data, use 1 p.e. = 90 g TSS/person/day.*
	- primary treatment capacity	p.e.	Installed *primary treatment* capacity *For this data, use 1 p.e. = 90 g TSS/person/day.*
	- secondary treatment capacity	p.e.	Installed *secondary treatment* capacity
	- tertiary treatment capacity	p.e.	Installed *tertiary treatment* capacity
	Peak flow storage capacity at WWTP	(m^3)	Maximum volume of wastewater that can be stored in *wastewater treatment plants*
	Pumping capacity at WWTP inlet	(kW)	Total nominal power of inlet pumps at *wastewater treatment plants*
TECHNOLOGICAL RESOURCES	Computerised information systems (IT=Information Technology)		Use of electronic equipment, especially computers, for storing, analysing and sending out information [10]
	- planning and decision	(yes/no)	Routine use of IT to support planning and decision
	- inspection	(yes/no)	Routine use of IT to support *inspection* activities (specify CCTV, when applicable)
	- maintenance	(yes/no)	Routine use of IT to support maintenance activities
	- customer *complaints*	(yes/no)	Routine use of IT to support customer service quality and network performance appraisal
	- other	(yes/no)	Routine use of IT to support other relevant management activities

10 *in* Oxford (2002).

CONTEXT INFORMATION	System profile data	(unit/option)	Concept *Comments*	variable
	Monitoring, automation and control		*In the items below, monitoring, automation and control devices availability is assessed.*	
	- flow meters	*(No.)*	Number of flowmeters installed in the wastewater system, either permanently or temporarily, during the year	
	- quality monitors	*(No.)*	Number of automatic samplers and online sensors used in the wastewater system, either permanently or temporarily, during the year	
	- pumping	*(%)*	Number of *pumping stations* with automatic control units / total number of pumping stations x 100	
	- treatment	*(yes/no)*	Routine use of automatic devices to control treatment units	
	- monitoring and control	*(yes/no)*	Availability and routine use of monitoring and control facilities (e.g. telemetry, SCADA, RTC)	
	- integrated control	*(%)*	Number of integrated stations / total number of *stations* x 100 *Integrated stations are stations with automatic control devices that are integrated in an automation network or are directly connected to a central control facility.*	
	Mapping			
	- updated mapping	*(%)*	Length of mapped sewers with a capture tolerance compatible to a scale 400 p.p.m. / total sewer length *400 p.p.m. = 1:2500*	
	- digital mapping	*(%)*	Length of digitised sewers with a capture tolerance compatible to a scale 400 p.p.m. / total sewer length *400 p.p.m. = 1:2500*	

8.4 Region profile

CONTEXT INFORMATION	Region profile data *(unit/option)*	Concept *Comments* — *variable*
DEMOGRAPHY AND ECONOMICS	Population density *(inhab./km^2)*	Resident population / area that is the responsibility of the wastewater undertaking
	Household occupancy *(inhab./house)*	Resident population / total number of *dwelling* units (houses + apartments)
	Population growth	*Based on official statistics, whenever possible*
	- current population growth rate *(% per year)*	(Population variation during the last ten years / population in first year of this period x 10) *In fast growing areas a 5-year period should be adopted.*
	- forecast population growth rate *(% per year)*	Forecast average yearly population growth rate for the future ten years
	Gross National Product per capita *(US$ per capita/year)*	Gross National Product / total country population *Use the official figure.*
	Inflation rate *(%/year)*	Annual percentage change in consumer price index in the country *Use the official figure.*
	US$ exchange rate	
	- annual average US$ exchange rate *(local currency/US$)*	Exchange ratio of the local currency for the US$, assessed as the average of the values at the end of each month, for the reference year *See section 7.7 - Economic and financial indicators (wFi) for US$ use justification.*
	- US$ exchange rate at the end of the assessment period *(local currency/US$)*	Exchange ratio of the local currency for the US$ at the end of the assessment period *See section 7.7 - Economic and financial indicators (wFi) for US$ use justification.*
	Unemployment rate *(%)*	Percentage of the labour force that actively seeks work but is unable to find work *Use the official figure. This context information considers persons of working age who, in the reference period, are without work, are available for work and have taken specific steps to find work.*
ENVIRONMENT	Yearly rainfall	*Provide the average figures over the past 30 years or the longest continuous period. These statistics relate to the area of service within the wastewater system area.*
	- average *(mm/year)*	Yearly average rainfall
	- maximum *(mm/year)*	Yearly maximum rainfall assessed as the annual maximum
	- minimum *(mm/year)*	Yearly minimum rainfall assessed as the annual minimum

CONTEXT INFORMATION	Region profile data *(unit/option)*	Concept *Comments* *variable*
	Short duration rainfall	*Based on the longest time series available*
	- availability of local statistical data *(yes/no)*	*If "Yes", answer the two following items*
	- 10 min, 10 year return period *(mm/year)*	Maximum intensity for 10 min duration and 10 year return period (value obtained from IDF curves)
	- 60 min, 10 year return period *(mm/year)*	Maximum intensity for 60 min duration and 10 year return period (value obtained from IDF curves)
	Air temperature	*Give the average over the past 30 years or the longest continuous period.*
	- daily average *(ºC)*	Daily average air temperature of the year
	- daily maximum *(ºC)*	Average air temperature for the hottest day of the year
	- daily minimum *(ºC)*	Average air temperature for the coldest day of the year
	Topography	
	- maximum altitude *(m)*	Maximum elevation above sea level within the wastewater system area that is the responsibility of the wastewater undertaking
	- minimum altitude *(m)*	Minimum elevation above sea level within the wastewater system area that is the responsibility of the wastewater undertaking
	Receiving bodies	
	- ocean *(yes/no)*	
	- estuaries, bays and other coastal waters *(yes/no)*	
	- rivers *(yes/no)*	
	- streams *(yes/no)*	
	- lakes, ponds, reservoirs or closed bays *(yes/no)*	
	- wetlands *(yes/no)*	
	- soil *(yes/no)*	
	Special protected areas *(yes/no)*	
	- special protected area *(km^2)*	If "yes", total area classified as *special protected area*.

9. REFERENCES

Alegre, H. (2002a). **Implementation of a performance indicators system**. Retrieved June 2002, from the workshop *Views and Experience Gained Through Implementing the IWA Performance Indicators Project* (Melbourne, Australia) Web site: www.iawq.org.uk/template.cfm?name=piworkshop

Alegre, H. (2002b). **Performance indicators as a management support tool**. *in* Mays, L.W., **Urban Water Supply Handbook**, MacGraw-Hill, ISBN 0-07-137160-5.

Alegre, H., Baptista, J.M., Lobato de Faria, A. (1997). **A general framework of performance indicators in the scope of water supply**. Workshop *Performance indicators for transmission and distribution systems*, Lisboa, Portugal.

Alegre, H., Hirner, W., Baptista, J.M., Parena, R. (2000). **Performance indicators for water supply services.** Manual of Best Practice Series, IWA Publishing, London, ISBN 1 900222 27 2.

Alegre, H., Hirner, W., Baptista, J.M., Parena, R. (2002a). **Highlights of the IWA system of performance indicators for water supply services**. Retrieved June 2002, from the workshop *Views and Experience Gained Through Implementing the IWA Performance Indicators Project* (Melbourne, Australia) Web site: www.iawq.org.uk/template.cfm?name=piworkshop

Alegre, H., Hirner, W., Baptista, J.M., Parena, R., Cubillo, F., Cabrera, E., Matos, R. (2002b). **The IWA Systems of Performance Indicators for Urban Water Services.** Retrieved June 2002, from the workshop *Views and Experience Gained Through Implementing the IWA Performance Indicators Project* (Melbourne, Australia) Web site: www.iawq.org.uk/template.cfm?name=piworkshop

Ashley, R M., Hopkinson, P. (2002). **Sewer systems and Performance indicators – into the 21st Century.** *Journal of Urban Water* **4** (2) pp. 123-136.

Ashley, R. M., Smith, H., Jowitt, P. H., Butler, D., Blackwood, D. J., Davies, J. W., Gilmour, D., Foxon, T. (2002). **Making More Sustainable Decisions For Asset Investment In The Water Industry – Sustainable Water Industry Asset Resource Decisions – The SWARD Project**. Proceedings of the *9th ICUD – International Conference on Urban Drainage*, Portland; Oregon, USA.

Ashley, R. M., Souter, N., Butler, D., Davies, J., Dunkerley, J., Hendry, S.(1999). **Assessment of the sustainability of alternatives for the disposal of domestic sanitary waste**. *Water Science and Technology* **39** (5) pp. 251- 258.

Audit Commission (UK) (2000). **On-target – the practice of performance indicators**. The Audit Commission, 1 Vincent Square, London SW1P 2PN, ISBN 1 86240 228 0, retrieved from the Audit Commission Web site: www.audit-commission.gov.uk

Deb A., Cesario L. (1997). **Water Distribution Systems Performance Assessment**. *IWSA Workshop on Performance Indicators for Transmission and Distribution Systems.* Lisbon, Portugal. Project funded by the AWWA Research Foundation.

Duarte, P., Alegre, H., Matos, R. (2003). **About the two IWA performance indicators systems for urban water management**, in *PEDS 2003 – Pumps, Electromechanical Devices and Systems Applied to Urban Water Management* Conference Proceedings, Valencia, Spain.

Ewan Associates Ltd / Mott MacDonald Ltd (2001). **Development of Enhanced Serviceability Indicators For Sewerage Assets**. Report ref. 01-G003/02/Ofwat/od/dg/002/63452.

Foxon, T. J., McIlkenny, G., Gilmour, D., Oltean-Dumbrava, C., Souter, N., Ashley, R., Butler, D., Pearson, P., Jowitt, P., Moir, J. (2001). **Sustainability Criteria For Decision Support In The UK Water Industry**. *J.Env. Planning & Management.* **45** (2) pp. 285-301, March.

Hopkinson, P., Whitaker, M. (1999). **The relationship between company environmental reports and their environmental performance: a study of the UK Water industry**. In: Bennet M., James P. (Eds.). *Sustainable Measures. Evaluating and reporting of environmental and social performance.* Sheffield: Greenleaf pub.

ISO/TC 224 (2003). **Service activities relating to drinking water supply systems and wastewater systems – quality criteria of the service and performance indicators**, Business Plan, Version: Draft 4, February 2003.

Larsson M., Parena R., Smeets E., Troquet I. (2002). **Process Benchmarking in the Water Industry**. Manual of Best Practice Series, IWA Publishing, London, ISBN 1 84339 010 8.

Matos, M.R., Cardoso, M.A, Duarte, P., Ashley, R., Molinari, A., Shulz, A. (2002a). **Performance Indicators for Wastewaters Services – Towards a Manual of Best Practice**, IWA 3rd World Water Congress, Melbourne, Australia.

Matos, M.R., Cardoso, M.A, Duarte, P., Ashley, R., Molinari, A., Shulz, A. (2002b). **The IWA performance indicator manual for wastewater service provision,** *in* International Conference on Sewer Operation and Maintenance Proceedings (SOM2002), Bradford, UK.

Matos, M.R., Cardoso, M.A, Duarte, P., Ashley, R., Molinari, A., Shulz, A. (2003). **Performance Indicators for Wastewaters Services – Towards a Manual of Best Practice**, *Water Science and Technology: Water Supply* **3** (1-2) pp. 365-371, IWA Publishing, London.

Merkel, W. (2002). **International Report: Performance assessment in the water industry**, *Water Science and Technology: Water Supply* **2** (4) pp. 151–162, IWA Publishing, London.

OFWAT (2002). **June return reporting requirements and definitions manual**. Retrieved March 2003, from the Office of Water Services (OFWAT) Web site: www.ofwat.gov.uk/aptrix/ofwat/ publish.nsf/Content/jr02_reporting_requirements

Oxford (2002). **Oxford Advanced Learner's Dictionary of Current English**, 6th edition, 2002 (8th impression), Oxford University Press, Oxford, UK. ISBN 0 19 431 585-1.

PBM SIG - Performance-Based Management Special Interest Group (2001). **The Performance-Based Management Handbook** - Vol. 1. Retrieved February 5, 2003, from Oak Ridge Institute for Science and Education Web site: http://www.orau.gov/pbm/pbmhandbook/ pbmhandbook.html

Reynolds J H. (2000). **Maintenance of sewers and water mains within the concept of serviceability to customers**. *Pipes & Pipelines International*. March-April.

Schulz, A., Schön, J. (2001). **A benchmarking system in the water-management industry from an operator's point of view - Requirements, using the example of treatment plants**. *IWA 2nd World Water Congress* Proceedings, Berlin, Germany.

Stahre P., Adamsson J. (2002). **Performance benchmarking: A Powerful Management Tool For Water And Wastewater Utilities.** Presented at AFNOR Atelier *Assurer la qualité de la gestion des services d'eau potable et d'assainissement: les bonnes pratiques – un enjeu, une méthodologie, un partage d'expériences*, Paris, France.

Turrell, R. (1995). **Levels of Service for the Municipal Services Sector.** Consultant report for the project *Sustainable development strategies for the water supply and wastewater sectors in Portugal*, developed for the General Directorate for the Environment (DGA, Portugal), funded by the Cohesion Fund of the European Commission.

UKWIR Ltd (2002). **Capital Maintenance Planning: A common framework.** Vol. 1 - Overview. 02/RG/05/3 ISBN 1-840572655.

WERF (1997). **Benchmarking wastewater operations – collection, treatment, and biosolids management**, Final Report Project 96-CTS-5. Water Environment Research Foundation, Alexandria, USA. ISBN 1-57278-121-1.

World Bank (2002). **Benchmarking Water & Sanitation Utilities: Network of Core Indicator Values.** Retrieved April 5, 2002, from World Bank Web site: www.worldbank.org/html/fpd/water/topics/ bench_network.html

10. ABOUT THE AUTHORS

Rafaela Matos

Laboratório Nacional de Engenharia Civil (LNEC) – DHA – NES
Av. do Brasil, 101
PT-1700-066 Lisboa, Portugal
Telephone (direct): + 351 21 844 36 26; Fax: + 351 21 844 30 32
E-mail: rmatos@lnec.pt

Born in Lourenço Marques, Mozambique, in 1954, graduated in civil engineering at the Lisbon Technical University, Portugal, in 1977. She became Researcher at the National Laboratory of Civil Engineering (LNEC) in 1987 with the thesis "*Methodologies for evaluation and analysis of urban stormwater - study based on experimental data in Portugal*". For her research work she was awarded the National Award for the Environment (1988), the Hydraulic and Water Resources Award (1988-1989), and the LNEC institutional prize, Manuel Rocha award (1992). In 2000 she was appointed to the top position of Principal Researcher. She has been deeply involved in Regulation and Standards, since 1985, being co-responsible for the draft projects for Regulation on Water Supply and Wastewater for Portugal (Europe), for Macao (South of China), for S. Tomé e Príncipe (Africa), and for Mozambique (Africa). She is the Portuguese delegate in the activities of the CEN/TC 165 - *Wastewater Engineering,* since 1991, and chaired the National Standard Technical Committee CT90 on Water Supply and Wastewater, between 1994 and 1999. She is Portuguese delegate and CEN/TC 165 official liaison at the ISO/TC224 – *Performance Indicators for water supply and wastewater services*, since 2002.

She is author or co-author of more than 200 scientific and technical publications and organised twenty national technical and scientific events and fifteen international events. Her field of expertise is urban drainage modelling and flooding control, sewer systems diagnosis and rehabilitation, urban water management, performance assessment of sewer systems. She has been the Head of the Water Supply and Wastewater Division of LNEC between 1990 and 2002, co-ordinating a team of 20 people dealing with applied research on the urban water cycle. She chaired the Portuguese Association for water supply, wastewater and solid wastes (APESB), between 1993 and 1999. Within the International Water Association, she is member of the Joint Committee on Urban Drainage (JCUD) since 2000. She is the leader of the IWA Task Force on "Performance Indicators for Wastewater Services ".

Adriana Cardoso

Laboratório Nacional de Engenharia Civil (LNEC) – DHA – NES
Av. do Brasil, 101
PT-1700-066 Lisboa, Portugal
Telephone (direct): + 351 21 844 3618; Fax: + 351 21 844 30 32
E-mail: macardoso@lnec.pt

Born in Lisbon, Portugal, in 1964, she graduated in civil engineering from the Lisbon Technical University, Portugal, in 1987. She received a MSc degree in Hydraulics and Water Resources from the Lisbon Technical University in 1992. She is a PhD student at the Lisbon Technical University and Research Assistant at the Urban Water Division of the National Laboratory of Civil Engineering (LNEC).

Her activity in the field of Urban Water has been mainly focused on urban drainage modelling, performance assessment of sewer systems and rehabilitation. She is author and co-author of more than 50 papers, articles, reports and thesis. She has been member of the Specialised Group on Water Supply and Sanitation of the Portuguese Association of Water Resources (APRH) since 1993 and she was General Secretariat of the Portuguese Association for Water Supply, Wastewater and Solid Wastes (APESB), between 1996 and 1999. She is member of the IWA Task Force on "Performance Indicators for Wastewater Services".

Richard Ashley

Pennine Water Group
School of Engineering, Design & Technology
University of Bradford
BD7 1DP West Yorkshire, UK
Telephone (direct): +44 1274 233865; Fax: +44 1274 233888
email r.ashley@bradford.ac.uk

Born on London in 1948, he graduated in civil engineering in 1972, whilst working for the Greater London Council rehabilitating main drainage in the capital. After a brief spell as a structural engineer, joined Thames Polytechnic as a hydraulics and public health engineering lecturer. Carried out research into wave loading on offshore structures with the National Maritime Institute and received a MPhil degree in 1981. Moved to Dundee Institute of Technology in 1982, becoming a chartered civil engineer (MICE) in 1988, after earlier qualifying as a member of the Institution of Water and Environmental Management. Chair of Water and Environmental Engineering at the University of Abertay Dundee and co-founder and Head of the Urban Water Technology Centre.

Moved to University of Bradford in 2000, as Professor of Urban Water, where he jointly established the Pennine Water Group (with the Unversity of Sheffield), a UK EPSRC funded platform grant centre of excellence. He is a visiting professor at the Technical University of Delft, a Committee member of the IWA/IAHR joint committee on Urban Drainage and founder and erstwhile Chairman of the Sewer Systems & Processes working group. He has published/presented some 200 papers in the urban drainage field and is chief editor and author of the forthcoming IWA scientific and technical report on Sewer Solids (state of the art). He is a contributor to the forthcoming IWA Glossary of Urban Drainage and contributor of a number of book chapters. He is the Chairman of the Yorkshire and Humberside Region of the Institution of Civil Engineers and working on two major UK projects dealing with the control of drainage flooding in response to climate change and the development of more sustainable water cycles management systems for new developments.

Patrícia Duarte

Laboratório Nacional de Engenharia Civil (LNEC) – DHA – NES
Av. do Brasil, 101
PT-1700-066 Lisboa, Portugal
Telephone (direct): + 351 21 844 3842; Fax: + 351 21 844 30 32;
E-mail: pduarte@lnec.pt

Born in Lisbon in 1973, she graduated in environmental engineering in 1996, from Universidade Nova de Lisboa. Since the beginning of 2003 she has been a PhD Student in Civil Engineering Hydraulics with thesis work on water supply distribution networks conception regarding technical performance optimisation. She worked at LNEC in the previous three years as a research trainee under a project concerning the rehabilitation of water supply and wastewater systems.

She is a member of the IWA Task Force on "Performance Indicators for Wastewater Services". She has also been a member of the IWA performance indicators for water supply services field-testing coordinating team and translated the IWA Water Supply Manual of Best Practices (Alegre *et al.*, 2000) into Portuguese. She has also been collaborating on the second edition of that Manual. Before her activity at LNEC, she worked in a private water supply and wastewater services management company for two years.

Alejo Molinari

Ente Tripartito de Obras y Servicios Sanitarios (ETOSS)
Av. Callao 982
(C1023 AAP) Buenos Aires, Argentina
Telephone / Fax:: +54 11 4816 5068;
E-mail: amolinari@etoss.org.ar

Born in Santa Rosa (L.P.), Argentina, in 1955, he graduated in Civil Engineering from the Southern National University (Argentina), in 1979, and from the Padua University, Italy, in 1980. He also has an MBA awarded by the Austral University (Argentina) in 1991 and other qualifications in Utilities Regulation and Strategy, from the PURC of Florida University (USA) and in Utilities Economic Regulation by the CEER of UADE University (Argentina). PhD in Civil Engineering.

He spent the earlier part of his career in the management of large infrastructure projects in Europe and the Middle East. In 1994 he became Asset Administration Manager for the Regulation Authority of Water and Sewage Services for the Buenos Aires Region (ETOSS). His duties covered the studies required for future investments, as well as in the predictive maintenance asset program. In 2000 he became the Service Quality Manager for the same Authority, being responsible for the quality of water, sewage and wastewater treatment services, coordinating a group of 11 specialists.

He is leading the Water Regulation Benchmarking Clearinghouse for the Argentine Association of Water Regulatory Authorities (AFERAS); and has been appointed as leader of the benchmarking project for the Association of Water and Sanitation Regulators of the Americas (ADERASA). Acting as a consultant, he has recently developed the regulatory benchmarking for the Bolivian water and sanitation Regulator Body (SISAB); and assisted the ULBRA University in Porto Alegre (Brazil) to set up a Master Degree in Public Services Strategy and Regulation.

He has organized several national and international events and has been a speaker at many seminars and international events on strategy, regulation and benchmarking of utilities.

Andreas Schulz

Emschergenossenschaft/Lippeverband
Kronprinzenstraße 25, 45128 Essen, Germany
Telephone: +49 (0) 201/104-2723; Fax: +49 (0) 201/104-2786
E-mail: aschulz@eglv.de

Born in 1956 in Wattenscheid, since incorporated into the city of Bochum. He studied Civil Engineering at the Technical University of Berlin, specializing in the design elements of hydraulic engineering in the field of "Design and agricultural hydraulic engineering", and graduated in 1981. He achieved his doctorate in 1986 with a thesis on the hydraulics of natural and artificial bodies of water and waterways. He has been with Emschergenossenschaft/Lippeverband since 1987, working firstly in the field of planning and construction of wastewater and rainwater treatment plants, then of canals and waterways, and most recently as head of the Construction Technology department. He was entrusted in 1997 with the setting-up of a Technical Supervision team in the Technical Division at Board level. Since mid 2001 he has been in charge of the department for Market and Customer. He is the coordinating manager for the development of benchmarking systems within the associations. He has been, and remains, an active member of a number of DVWK and ATV work groups, and is also the spokesman for the ad hoc "Benchmarking" team.

Appendix 1 – Glossary

TERMS RELATED TO THE WASTEWATER SYSTEM

ancillary: structures and units comprising the 'non-sewer' part of the *sewer system* (*pumping stations*, *storage tanks*, *CSOs*, etc.)

appropriate treatment: treatment of *municipal wastewater* by any process and/or disposal system that is deemed adequate for a specified objective.

blockage: caused by an obstruction that interferes with the movement of wastewater. Blockage is most often used to describe the effect from an unwanted flow obstruction (partial or full) in a pipe, channel, *combined sewer system*, *overflow* regulator or other *flow regulator*, outfall, gate, orifice, storm inlet, *screen* or other location where flow occurs (Ellis *et al.*, 2003). In the USA, *stoppage* is sometimes used to describe the results of a blockage that completely stops the flow.

BMPs: structural measures used to store or treat urban *stormwater runoff* to reduce *flooding*, remove pollution, and provide other amenities. Typical examples of BMPs include detention or retention facilities, *infiltration* facilities, wetlands, vegetative strips, filters, water quality inlets and others. (See also *source control*) (Ellis *et al.*, 2003).

calibration: adjusting and checking of a measuring instrument to ensure accurate readings (Lo, 1992); set of operations that establish, under specified conditions, the relationship between values of quantities indicated by a measuring instrument (or values represented by a material measure or from a representative model) and the corresponding values specified by standards (adapted from SP, 2002).

cleaning: range of techniques including flushing, manual excavation, swabbing, air scouring and pressure jetting intended to remove loose or soft deposits from inside sewer networks, *storage tanks* or other *storage* areas such as pump sumps (adapted from Alegre *et al.*, 2000).

collapse: critical failure of a sewer due to structural fabric disintegration (NASTT, 2001).

collected sewage: all effluent from properties and other waste-producing sources (such as industries) carried away by sewers during *dry weather* (*wet weather* excluded).

combined sewer overflow (CSO): device, on a combined or partlally *separate sewer system* or at a *sewage* treatment works that relieves the system of excess flow (EN 752-1, 1995). CSO structures may include or be associated with a chamber that has *storage*. (see also *overflow*).

combined sewer system or unitary system: sewer system designed to carry both wastewater and surface water in the same pipeline(s) (EN 752-1, 1995).

complaint: any customer complaint expressed to the service provider either personally (verbal or written), by telephone, fax, post, e-mail or any other format (adapted from Alegre *et al.*, 2000).

domestic sewage: see *sanitary sewage*.

domestic sewer: see *sanitary sewer*.

drain: pipeline, usually underground, designed to carry wastewater and/or surface water from a source to a sewer or other point of discharge (adapted from EN 752-1, 1995).

dry weather flow (DWF): rate of flow in a *drain* or *sewer system* in specified *dry weather* conditions (EN 752-1, 1995). Usually flow in a sewer when the previous dry weather period is at least 24 hours. The rate of *sewage* together with *infiltration*, within a sewer during dry weather (White *et al.*, 1997). IWEM, 1993 specifies as the average daily flow during seven consecutive days without rain (excluding a period which includes public or local holidays) following seven days during which the rainfall did not exceed 0.25mm on any one day.

dry weather: period during which there has not been any rainfall on the catchment.

dwelling: house, a flat or an apartment, where a person lives (Oxford, 2002)

exfiltration: escape of *sewage* from a *drain* or *sewer system* into surrounding ground (EN 752-1, 1995). The downward or outward leakage (or movement) of *runoff* through the periphery of a sewer or *infiltration* device into the underlying or enveloping soil layers. The term is also used to refer to the overspillage of runoff from a raised, marginal low-flow channel into an adjoining *storage* area as in a side-by-side infiltration basin design (Ellis *et al.*, 2003). The leakage or discharge of flows being carried by sewers into the ground through leaks in pipes, joints, manholes, or other sewer system structures (NASTT, 2001).

exported wastewater: wastewater removed from the *sewer system* or *wastewater treatment plants* by the undertaking and introduced to external systems not the responsibility of the undertaking.

flooding: condition where *sanitary sewage*, other wastewater and/or surface water escapes from or cannot enter a *drain* or *sewer system* and either lies on the surface or enters buildings (adapted from EN 752-1, 1995). See also *surface flooding.*

flow regulator (dynamic/static): flow control device, which is used to regulate the flow into or from a structure or system. Types of regulator may be classified as static or dynamic. Each type of flow regulator has a (non-linear) characteristic relationship between the *head* (water level) and the flow regulated *discharge* (output). This relationship is a function of the flow conditions upstream and downstream of the control device and whether or not the regulator is operating with a free discharge or under submerged conditions (Ellis *et al.*, 2003).

foul sewage: see *sanitary sewage.*

foul sewer: see *sanitary sewer.*

gully pot/chamber (inlet, inlet grade, catchbasin US): roadside or other chamber designed to collect *stormwater runoff* washed from a carriageway or other paved surface and designed to trap grit and usually litter prior to entry into the sewer or drainage network (Ellis *et al.*, 2003).

imported wastewater: wastewater introduced into the *sewer system* or *wastewater treatment plants* from external sources not the responsibility of the undertaking, previously authorised.

industrial wastewater (trade effluent): primarily the specific liquid wastes collected from industrial processing that has no further product recovery use. It contains various pollutants and toxic substances depending on the industry. It may contain small quantities of *domestic sewage* from sanitary facilities in the industrial plant (Ellis *et al.*, 2003).

infiltration: ingress of groundwater into a *drain* or *sewer system* (EN 752-1, 1995).

inflow: entry of surface water directly into the *sewer system* (White *et al.*, 1997). In the USA this is water not intended to be conveyed by the sewer. I.e. storm water into a *sanitary sewer.*

inspection: implementation of a formal procedure, as a rule written, the results of which are subsequently recorded in a form that enables the undertaking to assess the serviceability of the assets and to take corrective action as appropriate (Alegre *et al.*, 2000).

interceptor sewer (or intercepting sewer): large sewers that, in a combined system, convey the flow of wastewater to the treatment plant or other point of discharge. May also be used in separate or combined systems to collect the flows from main and *trunk sewers* and carry them to treatment points (adapted from EPA, 2003). Interceptor sewers are usually the largest and may have few connections.

intermittent overflow discharge: designed *overflow* or *spill* from a *CSO*, to protect downstream sewer from overloading. Strictly applicable to conditions during *wet weather* flows in a combined sewer, but these *overflows* may also occur in *dry weather* due to lack of downstream sewer capacity.

municipal wastewater: wastewater from municipal areas consisting predominantly of *domestic sewage.* In addition it may also contain storm water, *infiltration* water, trade or wastewater from industries located in the urban area. See also *sanitary sewage.*

on-site system: system whereby all the waste and wastewater produced is handled locally on the site near to point of origin, by means of *septic tanks*, soakaways, composting toilets, evaporation beds, reed beds etc. The waste may subsequently be utilised directly for fertiliser and/or used in energy recovery processes. Solid matter may be transported elsewhere and liquid effluents may discharge to main sewers. The term includes wet and dry systems.

overflow: usually related to the excess flows discharged from a *CSO*. Overflows may also occur due to malfunctioning sewer *flow regulators*. Also referred to as *spill.*

preliminary treatment: wastewater treatment process usually comprised of grit chambers and fine screening to remove relatively large grit, detritus, and floatable matter (Ellis *et al.*, 2003).

pre-treatment: treatment of wastewater prior to discharge into municipal collection systems (Metcalf, 1991)

primary treatment: wastewater treatment process that uses settling, skimming, and (optional) chlorination to remove suspended solids, floating materials, and pathogens from sanitary, combined, storm, and *industrial wastewater*. Primary treatment typically removes about 35% and 65% of 5-day biochemical oxygen demand (BOD_5) and SS, respectively. For higher suspended solids removal, chemicals (coagulants and polyelectrolytes) are added (into a coagulation chamber and the wastewater is then flocculated in another chamber) prior to the sedimentation tanks (adapted from Ellis *et al.*, 2003).

property: *dwelling* or other building, sub-part of building used for individual living or working for households, owners, tenants, visitors (if property is visited for any purpose), working or recreation. In apartment blocks, each separate household; in business properties, space occupied by individual businesses (Alegre *et al.*, 2000).

pump main: pipe through which *sewage* is pumped (EN 752-1, 1995). Also referred to as *rising main.*

pumping station: building, structures and equipment used to transfer *sewage* through a *rising main* or otherwise to lift or add hydraulic head to the sewage (adapted from EN 752-7, 1998).

receiving water: any body of water such as the sea, a river, stream or lake as well as an aquifer into which *drain* or *sewer systems* discharge (EN 752-1, 1995). A body of surface or subsurface water, which provides a dilution capacity for directly, discharged pollutants. Examples of surface receiving waters, which may be flowing or stationary, are streams, rivers, lakes, ponds, and tidal waters such as estuaries or the sea (see also *tidal waters*). The spatial and temporal distribution, dispersion and assimilation of pollutants within receiving waters are critical in determining their resultant impact (Ellis et al., 2003).

refurbishment: all methods for restoring existing assets in order to achieve the requisite performance (Alegre *et al.*, 2000). See also *rehabilitation.*

rehabilitation: all measures for restoring or upgrading the performance of existing *drain* and *sewer systems* (EN 752-1, 1995). All aspects of upgrading the performance of existing sewer systems. Structural rehabilitation includes *repair*, *renovation* and *renewal.* Hydraulic rehabilitation covers *replacement*, *reinforcement,* flow reduction or attenuation and occasionally renovation (WRc, 2001).

reinforcement: the provision of an additional sewer, which in conjunction with an existing sewer increases overall flow capacity (WRc, 2001).

relining: removal of all deposits from inside an existing sewer, followed by the *in situ* application of a non-structural lining to restore the internal surface and/or to provide corrosion protection. Techniques may utilise cement or epoxy mortar (relining is sometimes referred to as scraping and lining, *renovation* or reconditioning) (Alegre *et al.*, 2000).

renewal: construction of a new sewer, on or off the line of an existing sewer, the function and capacity of the new sewer being similar to those of the old (WRc, 2001).

renovation: work incorporating all or part of the original fabric of the *drain* or sewer by means of which its current performance is improved (EN 752-5, 1997).

repair: rectification of local damage (EN 752-5, 1997). Rectification of damage to the structural fabric of the sewer and the reconstruction of short lengths, but not the reconstruction of the whole of the pipeline (WRc, 2001).

replacement: construction of a new *drain* or sewer, on or off the line of an existing drain or sewer, the function of the new drain or sewer incorporating that of the old (EN 752-5, 1997). *renewal* is a particular form of replacement in which the function of the new facility is the same as that of the existing. In practice this usually means that it is of the same nominal diameter (for sewers), power (for pumping systems), etc. (adapted from Alegre *et al.*, 2000).

rising main: see *pump main.*

runoff: water from precipitation that flows off a surface to reach a *drain*, sewer or *receiving water* (EN 752-1, 1995).

sanitary flooding: flooding that has occurred from a separate *sanitary sewer* system by whatever cause.

sanitary sewage: discharges from domestic properties (kitchens, laundry rooms, lavatories, bathrooms, toilets and similar facilities) of water contaminated by use (adapted from EN 752-1, 1995). Also discharges from medical facilities such as hospitals. May include large solids introduced into WCs or other inlets. Also referred to as *domestic sewage* and *foul sewage*.

sanitary sewer overflow: structures designed to allow *overflows* from sanitary *sewer systems* when the downstream system reaches capacity. Similar to *CSOs*, except a component of a sanitary sewer system rather than *combined sewer system*.

sanitary sewer: sewer system designed to convey solely sanitary or foul flows, usually to end of pipe treatment plants. These sewers are designed to a capacity dependent upon user inputs to the system, with no allowance made for *stormwater* (other than *inflow* and *infiltration*). Flows in these pipes usually include some inflow and infiltration (adapted from Ellis *et al.*, 2003). Also referred to as *domestic sewer* and *foul sewer*.

screen: latticed metal grilles, (normally constructed of steel) intended to capture gross solids, litter and detritus in *stormwater* and wastewater flows at outfalls prior to discharge into *receiving waters* or used in conjunction with *CSO* structures or WWTP inlets. May be mechanically operated or static. (adapted from Ellis *et al.*, 2003).

secondary treatment: step in purifying wastewater for removing fine particulate, colloidal and dissolved organic (and inorganic) by using biological processes, including trickling fixed-film filtration and suspended growth (activated sludge). Recently, a sequential system providing aerobic, anoxic, and anaerobic environments is being used instead of the conventional activated sludge aeration process for effective biological transformation of toxic organic pollutants to less or non-toxic constituents. The mixed liquor flows from the aeration tank to the secondary sedimentation tank to remove the mixed liquor suspended solids. In a number of countries disinfection of the effluent prior to discharging into *receiving water* completes secondary treatment (adapted from Ellis *et al.*, 2003).

sediment: solid material that is conveyed from surfaces, properties, industry, in infiltrating water and other sources that is conveyed in the flow of *sewage* or *stormwater* and collects in sewers or *ancillaries* in regions of low velocity.

separate sewer system: sewer system, normally of two pipelines, one carrying *sanitary sewage* and the other surface water (adapted from EN 752-1, 1995).

septic tank: type of settlement tank intended to provide quiescent conditions for settlement of *sludge* and the development of anaerobic conditions for decomposition of organic matter. Raw *sewage* is fed to the tank, and settled sewage is discharged to the drainage field through the tank outlet. Sludge accumulates at the bottom of the tank and has to be removed periodically A septic tank system consists of the tank itself, and if it discharges to land, a drainage field (Payne *et al.*, 1993). See also *on-site system*.

service connection: part of the pipe system managed by the undertaking that conveys wastewater from a building to a common sewer (sewer that is the responsibility of the wastewater undertaking). Also called building sewer, house sewer (James, 1996). [Distinct from building or *property drain*].

service interruption: planned, unplanned (even if notified) or un-notified suspension of service to customers with a duration measured from start of suspension to full restoration of service. Caused by breaks, *collapses*, stoppages or failures in the *sewer system*, planned maintenance, construction, *repair/renewal* measures or whatever cause. Includes planned interruptions even where pre-notified to customers (adapted from Alegre *et al.*, 2000).

settled sewers: small-bore sewers that convey solids-free *sewage* usually from a *septic tank* or similar.

sewage: see *wastewater.*

sewer system (SE): network of pipelines and *ancillary* works which conveys *sanitary sewage* and/or surface water from *drains* to treatment works or other place of disposal (adapted from EN 752-1, 1995). Includes *interceptor sewers.*

sewer: pipeline or other construction, usually underground, designed to carry wastewater and/or surface water from more than one source

sewerage (SE): see *sewer system.*

shareholder: owner of shares in a company or business, in this case in a *sewerage* undertaker or other service provider (adapted from Oxford, 2002).

sludge disposal: process whereby *sludge* is removed from a treatment plant and dealt with elsewhere, i.e. by further treatment off-site, spreading on land, incineration, landfill etc. In some countries this may entail dumping at sea.

sludge stabilisation: usually anaerobic *sludge* digestion, a treatment that stabilizes raw sludge. Fully digested sludge has little readily biodegradable organic matter. It is not odorous and about 50% of the solids are inorganic. Sludge can also be digested aerobically (EPA, 2003).

sludge treatment: processing of wastewater *sludges* to render them innocuous. This may be done by aerobic or anaerobic digestion followed by drying in sand beds, filtering, and incineration, filtering, and drying, or wet air oxidation (EPA, 2003).

sludge utilisation: re-use of wastewater *sludge* usually following some form of conditioning, stabilisation or treatment . See also *sludge disposal.*

sludge: semi-solid residue produced from various wastewater treatment processes. Many types of sludge are classified as hazardous wastes and must be disposed of at a designated site (Ellis *et al.*, 2003). Sludge does not include grit and screenings collected during *preliminary treatment* of *domestic sewage* in a treatment works (EPA, 2003).

small-bore sewers: sewers of typically up to 150 mm diameter used to convey *sewage.*

source control: range of approaches and techniques for local, on-site management and control of *stormwater runoff* at the point of rainfall. Includes housekeeping procedures (non-structural), structural site controls and structural area or regional controls (adapted from Ellis *et al.*, 2003). May also apply to *on-site systems.*

special protected area: portions of land or waters (coastal or inland) protected by special restrictions and laws for the conservation of the natural environment. They include areas set aside for the protection of wildlife and its habitat; areas of great natural beauty or unique interest; areas containing rare forms of plant and animal life; areas representing unusual geologic formation; places of historic and prehistoric interest; areas containing ecosystems of special importance for scientific investigation and study; and areas which safeguard the needs of the biosphere (adapted from GEMET, 2002).

spill: see *overflow.*

stakeholder: any group or individual who can affect or is affected by the achievement of an organisation's objectives. Examples of stakeholders are employees, customers, suppliers, communities, individuals and even competitors (attributed to Edward Freeman, 1984 in Andriof *et al.*, 2002).

station: pumping stations, treatment plants and other control or monitoring points of the wastewater systems (adapted from Alegre *et al.*, 2000).

storage (inline or offline): use of a portion of the volume of a storm sewer or *drain*, *combined sewer system* and/or *interceptor sewer* system that is not being used to transport combined wastewater or *stormwater*, to accommodate the storage of additional stormwater *runoff* or sanitary wastewater. This term also applies to a storage facility, such as a tank, basin or other *reservoir* that is connected to a *sewer system* in such a way that all flow in the system passes through the storage facility. In the latter usage, inline storage is differentiated from offline storage, which is connected in such a way that excess flow can be diverted to the storage facility but normal flows bypass the facility (Ellis *et al.*, 2003).

storage tank / basin: reservoir intended to hold *stormwater*, combined, sanitary or *industrial wastewater* when temporary *storage* of the fluid is needed, typically when the flow rate or volume of flow from an event exceeds the carrying capacity of a pipe, channel, system or treatment facility, or when an allowable flow rate or total volume into a *receiving water* body is exceeded (Ellis *et al.*, 2003). See also *storage (inline or offline)*.

storm overflow device: structure that is built on a separate storm *sewer system* to *spill* to a *receiving water* course or a relief system, those excess flows that cannot be carried along the *sewer* (Ellis *et al.*, 2003). See also *combined sewer overflow (CSO)* and *sanitary sewer overflow*.

storm sewer overflow (SSO): see *combined sewer overflow (CSO)*. In some definitions, storm sewers may contain only *runoff* or also *sewage*, as in a combined sewer overflow (NetRegs, 2003).

stormwater: (1) water resulting from precipitation that either percolates into the soil, runs off freely from the surface, or is captured by a storm sewer, *combined sewer system*, and to a limited degree, *foul sewer* facilities. (2) surface water from rain, snow, or ice melting and running off from the surface of a drainage area. It is often collected in *sewers* separate from the *sanitary sewers*, and typically receives minimal, if any, treatment prior to discharge to *receiving waters* (James, 1996)

SUDS: sequence of management practices and control structures designed to drain surface water in a more sustainable fashion than some conventional techniques (may also be referred to as sustainable drainage techniques) (CIRIA, 2003).

surcharge: condition in which *wastewater* is held under pressure within a gravity *drain* or *sewer system*, but does not escape to the surface to cause *flooding* (adapted from EN 752-1, 1995).

surface flooding: condition where *wastewater* and/or surface water escapes from, or cannot enter, a *drain* or *sewer system* and either lies on the surface or enters buildings from the surface (EN 752-1, 1995). See also *flooding*.

tertiary treatment: advanced wastewater treatment that goes beyond the *secondary* biological treatment process. It removes a high percentage of SS (>95%), BOD_5 (>90%), and nutrients such as phosphorus and nitrogen. The tertiary processes include but are not limited to: nitrification and de-nitrification, sand and activated carbon filtration, and disinfection (Ellis *et al.*, 2003).

tidal waters: locations either on the land coast or offshore, where wastewater, partially treated wastewater or treated wastewater is discharged, and where there are tidal water movements.

treated wastewater in WWTP: wastewater that has received *primary*, *secondary*, or *tertiary treatment* to reduce its pollution or health hazards and is subsequently released as effluent from a wastewater facility after treatment (Horton, 2000).

trunk sewer: a *sewer* into which at least two branch sanitary, combined or *stormwater* sewers connect. It conveys the flow to the *interceptor sewers* (NASTT, 2001). See also *sewer system*.

wastewater reuse: wastewater that is used without first being discharged to a *receiving water*.

wastewater treatment plant (WWTP): series of treatment unit processes, including, but not limited to, one or more of: grit removal, screening, coagulation-flocculation, sedimentation, biooxidation, biofiltration, carbon adsorption, and disinfection by which pollutants are transformed and removed from wastewater streams. Treatment levels may be categorized as *preliminary* or pretreatment, *primary*, *secondary*, and *tertiary treatment*, depending on the degree of processing. Sewage treatment is the removal of pollutants or contaminants from sewage or wastewater (including sanitary, combined, and *stormwater*) for the protection of public health and the environment (adapted from Ellis *et al.*, 2003). Often abbreviated WWTP.

wastewater undertaking: the provider of the services that operates and manages the wastewater system under consideration.

wastewater: water changed by use and discharged to a *drain* or *sewer system* (EN 752-1, 1995). Also referred to as *sewage*.

wet weather: period when there is rainfall. Usually relates to rainfall sufficient to produce *runoff*. See also *dry weather*.

wrong connection: a connection into a *stormwater* sewer, or other stormwater conveyance system, such as an open channel, from a *sewer* that conveys *sewage* [either a house *drain*, *foul sewer*, or a combined sewer]. A connection into a *sanitary sewer* from a stormwater source [may or may not be a stormwater sewer].

TERMS RELATED TO POPULATION

commercial p.e.: part of the *service population equivalent in the area* related to the *wastewater* produced by commerce and restaurants operating in the area that is the responsibility of the undertaking.

imported wastewater population equivalent: population equivalent to the *imported wastewater*, comprising *wastewater* produced by any sources (population, industry, commerce, services and other) located out of the area that is the responsibility of the undertaking.

industry p.e.: part of the *service population equivalent in the area* related to the *wastewater* produced in the industries working in the area that is the responsibility of the undertaking, excluding those with authonomous *wastewater treatment plants* that discharge their treated wastewater directly to receiving bodies

other p.e.: part of the *service population equivalent in the area* related to *wastewater* produced by other sources than residential, industrial, commercial or services.

peak population: maximum numbers in the resident and non-resident population in the area that is the responsibility of the wastewater undertaking

population equivalent (p.e.): quantitative expression of the pollution load of commercial or *industrial wastewater* in terms of the number of "equivalent" people that would create a waste of the same strength. 1 p.e. corresponds to the pollution load of *sewage* generated by one inhabitant. Usually it corresponds to BOD_5 loads, but it can also be estimated from the TSS, COD or other per capita values. In this document, 1 p.e. = 60 g BOD_5/day equivalence shall be used by default while assessing variables, performance indicators and context information, except in the cases where other equivalences are explicitly defined.

population equivalent not served in the area: part of the *service population equivalent in the area* for which *wastewater* is neither collected nor treated by facilities that are the responsibility of the wastewater undertaking. It does not include wastewater treated by *on-site systems* and local systems (such as private *wastewater treatment plants* for industries and *septic tanks* in isolated areas) that are not the responsibility of the wastewater undertaking.

population equivalent served by on-site systems: part of the *service population equivalent in the area* for which *wastewater* is treated by on-site facilities (such as *septic tanks* and latrines). Some of these facilities provide a wastewater treatment solutions that conform with performance criteria, whereas other facilities may not satisfy the criteria. Although discharges from on-site WWT facilities may enter the sewer network, *interceptor sewers* or *wastewater treatment plants*, this is not considered good pratice, but may occur and is represented in the PI scheme by dashed arrows.

population equivalent served by sewer system and interceptors: part of the *service population equivalent in the area* and portion of the *imported wastewater population equivalent* for which *wastewater* is collected by *sewer systems* and interceptors that are the responsibility of the wastewater undertaking.

population equivalent served by the wastewater undertaking: part of the *service population equivalent in the area* for which *wastewater* is collected or treated by facilities that are the wastewater undertaking responsibility and *imported wastewater population equivalent.*

population equivalent served by wastewater treatment plants: part of the *service population equivalent in the area* and portion of the *imported wastewater population equivalent* for which *wastewater* is treated in *wastewater treatment plants* that are the responsibility of the wastewater undertaking.

population equivalent with satisfactory wastewater treatment: population equivalent that is served by *wastewater treatment plants* complying with effluent discharge consents.

population: number of inhabitants of a settlement area normally resident in that area (excludes residents with second homes in the area)

resident population: part of the *service population equivalent in the area* related to population living permanently in the area that is the responsibility of the undertaking.

seasonal population equivalent: part of the *service population equivalent in the area* related to the production of *wastewater* during only part of the period, such as from a seasonal population and seasonal industries. It corresponds to the following sum: seasonal population (number of persons x stay duration / assessment period) + seasonal industries (p.e. x operation duration / assessment period)

served service population equivalent in the area: part of the *service population equivalent in the area* which is served by collection or treatment facilities operated by the wastewater undertaking.

service population equivalent in the area: population equivalent to all the *wastewater* produced in the geographical area that is the responsibility of the undertaking and for whom a collection and/or treatment service should be provided by the wastewater undertaking. Wastewater that is treated by other autonomous solutions (such as private *wastewater treatment plants* in industries and on-site, e.g. *septic tanks* existing in isolated areas) shall not be accounted for.

services p.e.: part of the *service population equivalent in the area* related to the *wastewater* produced by offices, hospitals, schools, and other services that functioned in the area that is the responsibility of the undertaking.

REFERENCES

Alegre, H., Hirner, W., Baptista, J.M., Parena, R. (2000). **Performance indicators for water supply services.** Manual of Best Practice Series, IWA Publishing, London, ISBN 1 900222 27 2.

Andriof, J., Waddock, S., Husted, B., Rahman, S.S. (Eds.) (2002). **Unfolding stakeholder thinking.** Greenleaf pub. ISBN 1 87471 952 7.

CIRIA (2003). **Sustainable drainage systems – Glossary.** Retrieved March 2003, from Construction Industry Research and Information Association (CIRIA) Web site: http://www.ciria.org/suds/ glossary.htm

Ellis, J.B., Marsalek, J., Chocat, B., Fujita, S., Rauch, W. (2003). **Urban drainage glossary.** IWA Publishing, London (to be published). Draft version dated September 2001.

EN 752-1 (1995). **Drain and sewer systems outside the buildings – Part 1: Generalities and definitions.** European Committee for Standardisation (CEN).

EN 752-5 (1997). **Drain and sewer systems outside the buildings – Part 5: Rehabilitation.** European Committee for Standardisation (CEN).

EN 752-7 (1998). **Drain and sewer systems outside the buildings – Part 7: Maintenance and Operations.** European Committee for Standardisation (CEN).

EPA (2003). **Terminology Reference System.** Retrieved January 2003, from the U. S. Environmental Protection Agency Web site: http://www.epa.gov/trs/index.htm

GEMET (2002). **General Multilingual Environmental Thesaurus (GEMET).** European Environment Agency (EEA), European Topic Centre on Catalogue of Data Sources (ETC/CDS). Retrieved January 2003 from GEMET Web site: http://www.mu.niedersachsen.de/cds/etc-cds_neu/library/data/gemet_ENG.html? language=ENG&thesaurus=1

Horton, G. A. (2000). **Water Words Dictionary - A Compilation of Technical Water, Water Quality, Environmental, and Water-Related Terms.** Retrieved October 2002, from Nevada Division of Water Resources - Department of Conservation and Natural Resources Web site: http://water.nv.gov/Water%20planning/dict-1/ww-index.htm

IWEM (1993). **Handbooks of UK wastewater practice - Glossary.** Institution of Water and Environmental Management, London, UK.

James, W. (1996). **Hydraulics and hydrology vocabulary, Web Manual.** Retrieved October 2002, from the University of Guelph - School of Engineering Web site: http://www.eos.uoguelph.ca/webfiles/james/homepage/stillunclassified/wjvocab.html

Lo, S.S. (1992). **Glossary of Hydrology.** Water Resources Publications, Littleton, Colorado, USA.

Metcalf & Eddy (1991), **Wastewater Engineering – Treatment, disposal and reuse.** McGraw-Hill, 3rd Edition.

NASTT (2001) **Glossary of terms.** Retrieved July, 2002 from the North American Society for Trenchless Technology (NASTT) Web site: http://www.nastt.org/glossary/a.html

NetRegs (2003). NetRegs Glossary. Retrieved March, 2003 from the NetRegs Development Web site: http://www.environment-agency.gov.uk/netregs/help/275530/?version=1&lang=_e

Oxford (2002). **Oxford Advanced Learner's Dictionary of Current English**, 6th edition, 2002 (8th impression), Oxford University Press, Oxford, UK. ISBN 0 19 431 585-1.

Payne J A., Butler D. (1993). **Septic tanks and small sewage treatment works.** CIRIA Technical note 146, London, ISBN 0 86017 374 7.

SP (2002). **Metrology – Terminology.** Retrieved February 2003, from SP - Swedish National Testing and Research Institute Web site: http://www.sp.se/metrology/eng/terminology.htm

White, M., Johnson, H., Anderson, G., Misstear, B. (1997). **Control of infiltration to sewers.** CIRIA Report 175, London, ISBN 0 86017 474 3.

WRc (2001). **Sewerage Rehabilitation Manual – Fourth Edition, Volume 2.** WRc plc., UK, ISBN 1 898920 39 7.

Appendix 2 – Variables definitions

VARIABLES LIST

ENVIRONMENTAL DATA

ancillaries sediments	wA21	*wEn13*
Daily peak *preliminary* treated wastewater	wA4	*wPh1*
Daily peak *primary* treated wastewater	wA6	*wPh2*
Daily peak *secondary* treated wastewater	wA8	*wPh3*
Daily peak *tertiary* treated wastewater	wA10	*wPh4*
Number of *overflow* discharges	wA24	*wEn3*
on-site system sediments	wA23	*wEn15*
Other *sludge disposal*	wA19	*wEn8(wA16), wEn11*
population equivalent with satisfactory wastewater treatment	wA1	*wEn1*
preliminary treated wastewater	wA3	*wEn2(wA2), wQS5(wA2), wQS6*
primary treated wastewater	wA5	*wEn2(wA2), wQS5(wA2), wQS7*
Rainfall volume	wA26	*wEn5*
Screenings and grit	wA22	*wEn14*
secondary treated wastewater	wA7	*wEn2(wA2), wQS5(wA2), wQS8*
sewer sediments	wA20	*wEn12*
sludge disposed	wA16	*wEn8*
sludge going to landfill	wA17	*wEn8(wA16), wEn9*
sludge handled	wA14	*wEn7, wEn8, wEn9, wEn10, wEn11*
sludge produced in WWTP	wA13	*wEn6*
sludge thermally processed	wA18	*wEn8(wA16), wEn10*
sludge utilised	wA15	*wEn7*
tertiary treated wastewater	wA9	*wEn2(wA2), wQS5(wA2), wQS9*
Volume of *overflow* discharges	wA25	*wEn4, wEn5*
Wastewater reused	wA12	*wEn2*
Wastewater treated	wA2	*wEn2, wQS5*
Wastewater treated by *on-site systems*	wA11	*wEn2(wA2), wQS5*

PERSONNEL DATA

Absenteeism	wB23	*wPe22, wPe24(wB25)*
Absenteeism due to accidents or illness at work	wB24	*wPe23, wPe24(wB25)*
Absenteeism due to other reasons	wB25	*wPe24*
Basic education personnel	wB16	*wPe15, wPe16(wB17)*
Confined spaces skilled personnel	wB18	*wPe19*
Customer service personnel	wB7	*wPe6*
Financial and commercial services personnel	wB6	*wPe5*
General management personnel	wB4	*wPe3*
Human resources management personnel	wB5	*wPe4*
Laboratory personnel	wB13	*wPe7(wB8), wPe12, wPe13(wB8)*
Normal work	wB26	*wPe25*
Operation and maintenance personnel	wB12	*wPe7(wB8), wPe9, wPe13(wB8), wPe18, wPe19, wOp55*
Other qualification personnel	wB17	*wPe16*
Overtime work	wB27	*wPe25*
Planning, design and construction personnel	wB11	*wPe7(wB8), wPe8, wPe13(wB8)*
sewer system personnel	wB3	*wPe2*
Support services personnel	wB14	*wPe13*
Technical SE personnel	wB10	*wPe7(wB8), wPe11, wPe13(wB8)*
Technical services personnel	wB8	*wPe7, wPe13*

Technical WWT personnel	wB9	*wPe7(wB8), wPe10, wPe13(wB8)*
Total personnel	wB1	*wPe3, wPe4, wPe5, wPe6, wPe7, wPe8, wPe9, wPe14, wPe15, wPe16, wPe17, wPe20, wPe21, wPe22, wPe23, wPe24*
Training time	wB20	*wPe17*
University degree personnel	wB15	*wPe14, wPe16(wB17)*
Vaccinated personnel	wB19	*wPe18*
Working accidents	wB21	*wPe20*
Working fatalities	wB22	*wPe21*
WWT personnel	wB2	*wPe1*

PHYSICAL ASSETS DATA

Automated control units	wC17	*wPh11*
Connected properties	wC28	*wQS10, wQS11, wQS12, wQS13, wQS14, wQS15*
Control units	wC16	*wPh11, wPh12*
Daily *preliminary treatment* capacity	wC12	*wPh1*
Daily *primary treatment* capacity	wC13	*wPh2*
Daily *secondary treatment* capacity	wC14	*wPh3*
Daily *tertiary treatment* capacity	wC15	*wPh4*
Electrical switchgear	wC32	*wOp17*
Emergency power systems	wC30	*wOp15*
gully pots	wC22	*wOp4, wOp5*
Highly *surcharged sewers*	wC4	*wPh7*
Manhole chambers	wC21	*wOp3, wOp25, wOp26*
Number of *CSOs*	wC20	*wOp6, wOp43*
Number of *storage tank*	wC23	*wOp6*
overflow devices	wC19	*wEn3, wEn4*
Remote control units	wC18	*wPh12*
screens of tanks and *CSOs*	wC25	*wOp9*
SE flowmeters	wC27	*wOp12*
service connections	wC29	*wOp27*
sewer system pumping stations	wC9	*wPh10*
sewer system pumping stations utilisation	wC11	*wPh10*
sewer system pumps power	wC10	*wPh8, wOp11*
Signal transmission equipment	wC31	*wOp16*
Surcharged *sewers* in *dry weather*	wC2	*wPh5*
Surcharged *sewers* in *wet weather*	wC3	*wPh6*
Total *sewer* length	wC1	*wEn12, wEn13, wEn14, wPe2, wPe11, wPh5, wPh6, wPh7, wOp1, wOp2, wOp21, wOp22, wOp23, wOp24, wOp31, wOp32, wOp33, wOp34, wOp35, wOp37, wOp38, wOp39, wOp40, wOp54, wFi6, wFi8, wFi10*
Volume of tanks and *CSOs*	wC24	*wOp7, wOp8*
Wastewater quality monitoring instruments	wC33	*wOp14*
Wastewater system *pumping stations*	wC6	*wOp10, wOp36, wOp42*
Wastewater system pumps	wC7	*wOp41*
Wastewater system pumps power	wC8	*wOp28, wOp29*
WWTP flowmeters	wC26	*wOp13*
WWTP pumps power	wC5	*wPh9*

OPERATIONAL DATA

BOD tests carried out	wD48	*wOp44(wD46), wOp45*
BOD tests required	wD58	*wOp44(wD57), wOp45*
COD tests carried out	wD49	*wOp44(wD46), wOp46*
COD tests required	wD59	*wOp44(wD57), wOp46*
Co-generated energy	wD17	*wOp19*
CSOs controlled	wD45	*wOp43*
Electrical switchgear *inspection*	wD24	*wOp17*
Emergency power systems units *inspection*	wD22	*wOp15*
exfiltration volume	wD37	*wOp30, wOp33*
Faecal *E.coli* tests carried out	wD53	*wOp44(wD46), wOp50*
Faecal *E.coli* tests required	wD63	*wOp44(wD57), wOp50*
floodings from combined sewers	wD42	*wOp38*
floodings from *sanitary sewers*	wD41	*wOp37*
Gas detectors	wD68	*wOp55, wOp56*
gully pots cleaning	wD5	*wOp5*
gully pots inspection	wD4	*wOp4*
Industrial discharges tests carried out	wD56	*wOp53*
Industrial discharges tests required	wD66	*wOp53*
infiltration volume	wD36	*wOp30, wOp32*
inflow volume	wD35	*wOp30, wOp31*
Manhole chamber *inspection*	wD3	*wOp3*
Manhole chamber *replacement, renewal, renovation* or *repair*	wD29	*wOp25*
Manhole covers *replacement*	wD30	*wOp26*
Other wastewater quality tests carried out	wD54	*wOp44(wD46), wOp51*
Other wastewater quality tests required	wD64	*wOp44(wD57), wOp51*
Permanent vehicles	wD67	*wOp54*
Permanently installed gas detectors	wD69	*wOp56*
Power failures	wD18	*wOp42*
Pump energy consumption	wD12	*wOp20*
Pump energy consumption in *sewer system*	wD15	*wPh8, wOp20(wD12)*
Pump energy consumption in WWTP	wD14	*wPh9, wOp20(wD12)*
Pump *inspection*	wD11	*wOp11*
Pump refurbisment	wD32	*wOp28*
Pump *replacement*	wD33	*wOp29*
pumping station blockages	wD40	*wOp36*
pumping station inspection	wD10	*wOp10*
Pumps out of order	wD34	*wOp41*
screen maintenance in tanks and *CSOs*	wD9	*wOp9*
SE flow meter *calibrations*	wD20	*wOp12*
service connection replacement or *renewal*	wD31	*wOp27*
sewer and joints *repair*	wD28	*wOp24*
Sewer *blockage* locations	wD39	*wOp35*
Sewer *blockages*	wD38	*wOp34*
Sewer *cleaning*	wD2	*wOp2*
Sewer *collapses*	wD44	*wOp40*
Sewer *inspection*	wD1	*wOp1*
Sewer *rehabilitation*	wD25	*wOp21*
Sewer *renovation*	wD26	*wOp22*
Sewer *replacement*	wD27	*wOp23*
Signal transmission units *inspection*	wD23	*wOp16*
sludge tests carried out	wD55	*wOp52*
sludge tests required	wD65	*wOp52*
Standardisation factor	wD16	*wOp20*
storage tanks and *CSOs inspection* number	wD6	*wOp6*
surface floodings	wD43	*wOp39*
Tanks and *CSOs cleaning*	wD8	*wOp8*
Tanks and *CSOs inspected* volume	wD7	*wOp7*

Total nitrogen tests carried out	wD52	*wOp44(wD46), wOp49*
Total nitrogen tests required	wD62	*wOp44(wD57), wOp49*
Total phosphorus tests carried out	wD51	*wOp44(wD46), wOp48*
Total phosphorus tests required	wD61	*wOp44(wD57), wOp48*
TSS tests carried out	wD50	*wOp44(wD46), wOp47*
TSS tests required	wD60	*wOp44(wD57), wOp47*
Wastewater quality monitoring instruments *calibrations*	wD21	*wOp14*
Wastewater quality tests carried out	wD46	*wOp44*
Wastewater quality tests carried out by undertaking laboratories	wD47	*wPe12*
Wastewater quality tests required	wD57	*wOp44*
WWT energy consumption	wD13	*wOp18, wOp19*
WWTP flow meter *calibrations*	wD19	*wOp13*

DEMOGRAPHY (AND CUSTOMER) DATA

population equivalent served by *on-site systems*	wE6	*wEn15, wPe1, wPe10*
population equivalent served by WWTP	wE5	*wEn1, wEn6, wPe1, wPe10, wOp18*
Resident population	wE1	*wQS1, wQS2, wQS3, wQS4, wQS19, wQS20, wQS21, wQS22, wQS23, wQS24, wQS25, wQS26*
Resident population connected to SE	wE4	*wQS1*
Resident population not served	wE7	*wQS4*
Resident population served by *on-site systems*	wE3	*wQS3*
Resident population served by WWTP	wE2	*wQS2*
Total *population equivalent* served	wE8	*wFi1, wFi5, wFi7, wFi9, wFi27*

QUALITY OF SERVICE DATA

Accidents	wF22	*wQS28*
blockage complaints	wF13	*wQS19(wF12), wQS20, wQS27(wF12)*
collected sewage	wF1	*wOp30, wQS5, wQS6, wQS7, wQS8, wQS9*
Connections *repair* time	wF11	*wQS17*
Connections *repaired*	wF10	*wQS17*
Customer account related *complaints*	wF18	*wQS19(wF12), wQS25, wQS27(wF12)*
dry weather flooding of properties from combined sewers	wF4	*wQS12*
dry weather flooding of properties from *sanitary sewers*	wF2	*wQS10*
flooding complaints	wF14	*wQS19(wF12), wQS21, wQS27(wF12)*
New connections established	wF8	*wQS16*
New *service connections* establishment time	wF9	*wQS16*
Odour *complaints*	wF16	*wQS19(wF12), wQS23, wQS27(wF12)*
Other *complaints*	wF19	*wQS19(wF12), wQS26, wQS27(wF12)*
Pollution *complaints*	wF15	*wQS19(wF12), wQS22, wQS27(wF12)*
Responses to *complaints*	wF20	*wQS27*
Rodents related *complaints*	wF17	*wQS19(wF12), wQS24, wQS27(wF12)*
septic tank emptying requests	wF23	*wQS18*
septic tank emptying response time	wF24	*wQS18*
Third party damages	wF21	*wQS28*
Total *complaints*	wF12	*wQS19, wQS27*
Traffic disturbances	wF25	*wQS29*
Traffic lane interruptions	wF26	*wQS29*
Wastewater *service interruptions*	wF7	*wQS15*
wet weather flooding of properties from combined sewers	wF5	*wQS13*
wet weather flooding of properties from *sanitary sewers*	wF3	*wQS11*
wet weather surface flooding of properties	wF6	*wQS14*

FINANCIAL DATA

Accounts receivable	wG34	*wFi32*
Annual amount billed per year	wG40	*wFi37*
Annual debt from customers	wG39	*wFi37*
Capital costs	wG7	*wFi5, wFi6(wG5), wFi9, wFi10, wFi25, wFi26, wFi30(wG5)*
Capitalised costs of self-constructed assets	wG33	*wFi1, wFi2(wG1), wFi3(wG1), wFi30(wG1), wFi31(wG1)*
Cash-flow	wG41	*wFi39*
Current assets	wG45	*wFi41*
Current liabilities	wG49	*wFi40(wG43), wFi41, wFi43(wG44)*
Customer service management running costs	wG20	*wFi19*
Depreciated historical value of tangible assets	wG37	*wFi35, wFi42*
Depreciation costs	wG26	*wFi5(wG7), wFi6(wG5), wFi9(wG7), wFi10(wG7), wFi25, wFi26(wG7), wFi30(wG5), wFi33, wFi36*
Energy costs	wG11	*wFi5(wG6), wFi6(wG5), wFi7(wG6), wFi8(wG6), wFi11(wG6), wFi12(wG6), wFi13, wFi14(wG6), wFi15(wG6), wFi16(wG6), wFi17(wG6), wFi18(wG6), wFi19(wG6), wFi20(wG6), wFi21(wG6), wFi22(wG6), wFi23(wG6), wFi24(wG6), wFi30(wG5), wFi31(wG6)*
Exceptional earnings and losses	wG15	*wFi5(wG6), wFi6(wG5), wFi7(wG6), wFi8(wG6), wFi11(wG6), wFi12(wG6), wFi13(wG6), wFi14(wG6), wFi15, wFi16(wG6), wFi17(wG6), wFi18(wG6), wFi19(wG6), wFi20(wG6), wFi21(wG6), wFi22(wG6), wFi23(wG6), wFi24(wG6), wFi30(wG5), wFi31(wG6)*
External services costs	wG10	*wFi5(wG6), wFi6(wG5), wFi7(wG6), wFi8(wG6), wFi11(wG6), wFi12, wFi13(wG6), wFi14(wG6), wFi15(wG6), wFi16(wG6), wFi17(wG6), wFi18(wG6), wFi19(wG6), wFi20(wG6), wFi21(wG6), wFi22(wG6), wFi23(wG6), wFi24(wG6), wFi30(wG5), wFi31(wG6)*
Financial and commercial management running costs	wG19	*wFi18*
Financial debt service	wG42	*wFi39*
General management running costs	wG17	*wFi16*
Historical value of tangible assets	wG38	*wFi35, wFi36, wFi42*
Human resources management running costs	wG18	*wFi17*
Industrial revenues	wG4	*wFi4*
Interest expenses costs	wG27	*wFi5(wG7), wFi6(wG5), wFi9(wG7), wFi10(wG7), wFi25(wG7), wFi26, wFi30(wG5)*
Interest Income	wG28	*wFi5(wG7), wFi6(wG5), wFi9(wG7), wFi10(wG7), wFi25(wG7), wFi26, wFi30(wG5)*
Internal manpower costs	wG9	*wFi5(wG6), wFi6(wG5), wFi7(wG6), wFi8(wG6), wFi11, wFi12(wG6), wFi13(wG6), wFi14(wG6), wFi15(wG6), wFi16(wG6), wFi17(wG6), wFi18(wG6), wFi19(wG6), wFi20(wG6), wFi21(wG6), wFi22(wG6), wFi23(wG6), wFi24(wG6), wFi30(wG5), wFi31(wG6)*
Inventories	wG47	*wFi38*
Investment in tangible assets	wG30	*wFi27, wFi28, wFi29, wFi34*
Investments financed by the cash-flow	wG36	*wFi34*
Investments for assets *replacement* and *renovation*	wG32	*wFi27(wG30), wFi28(wG30), wFi29, wFi34(wG30)*
Investments for new assets and *reinforcement* of existing assets	wG31	*wFi27(wG30), wFi28, wFi29(wG30), wFi34(wG30)*
Investments subject to depreciation	wG35	*wFi33*
Leasing and rentals	wG13	*wFi5(wG6), wFi6(wG5), wFi7(wG6), wFi8(wG6), wFi11(wG6), wFi12(wG6), wFi13(wG6), wFi14(wG6), wFi15, wFi16(wG6), wFi17(wG6), wFi18(wG6), wFi19(wG6), wFi20(wG6), wFi21(wG6), wFi22(wG6), wFi23(wG6), wFi24(wG6), wFi30(wG5), wFi31(wG6)*
Long term liabilities	wG48	*wFi40(wG43), wFi43(wG44)*
Net income	wG52	*wFi43*
Net interest	wG29	*wFi5(wG7), wFi6(wG5), wFi9(wG7), wFi10(wG7), wFi25(wG7), wFi26, wFi30(wG5)*
Operating income	wG50	*wFi42, wFi44*
Operating income related taxes	wG51	*wFi44*

Operational costs	wG8	*wFi5(wG6), wFi6(wG5), wFi7(wG6), wFi8(wG6), wFi11(wG6), wFi12(wG6), wFi13(wG6), wFi14(wG6), wFi15(wG6), wFi16(wG6), wFi17(wG6), wFi18(wG6), wFi19(wG6), wFi20(wG6), wFi21(wG6), wFi22(wG6), wFi23(wG6), wFi24(wG6), wFi30(wG5), wFi31(wG6)*
Other operating costs	wG16	*wFi5(wG6), wFi6(wG5), wFi7(wG6), wFi8(wG6), wFi11(wG6), wFi12(wG6), wFi13(wG6), wFi14(wG6), wFi15, wFi16(wG6), wFi17(wG6), wFi18(wG6), wFi19(wG6), wFi20(wG6), wFi21(wG6), wFi22(wG6), wFi23(wG6), wFi24(wG6), wFi30(wG5), wFi31(wG6)*
Purchased merchandises	wG12	*wFi5(wG6), wFi6(wG5), wFi7(wG6), wFi8(wG6), wFi11(wG6), wFi12(wG6), wFi13(wG6), wFi14, wFi15(wG6), wFi16(wG6), wFi17(wG6), wFi18(wG6), wFi19(wG6), wFi20(wG6), wFi21(wG6), wFi22(wG6), wFi23(wG6), wFi24(wG6), wFi30(wG5), wFi31(wG6)*
Running costs	wG6	*wFi5, wFi6(wG5), wFi7, wFi8, wFi11, wFi12, wFi13, wFi14, wFi15, wFi16, wFi17, wFi18, wFi19, wFi20, wFi21, wFi22, wFi23, wFi24, wFi30(wG5), wFi31*
Service revenues	wG3	*wFi2, wFi3, wFi4, wFi32, wFi45*
sewer system running costs	wG23	*wFi22*
shareholders' equity	wG44	*wFi40, wFi43*
Support services running costs	wG25	*wFi24*
Taxes, levies and fees	wG14	*wFi5(wG6), wFi6(wG5), wFi7(wG6), wFi8(wG6), wFi11(wG6), wFi12(wG6), wFi13(wG6), wFi14(wG6), wFi15, wFi16(wG6), wFi17(wG6), wFi18(wG6), wFi19(wG6), wFi20(wG6), wFi21(wG6), wFi22(wG6), wFi23(wG6), wFi24(wG6), wFi30(wG5), wFi31(wG6)*
Technical management costs	wG21	*wFi20*
Total assets	wG46	*wFi40(wG44), wFi43(wG44), wFi44, wFi45*
Total costs	wG5	*wFi5, wFi6, wFi30*
Total debt	wG43	*wFi40, wFi43(wG44)*
Total operating revenues	wG2	*wFi1, wFi2(wG1), wFi3(wG1), wFi30(wG1), wFi31(wG1), wFi38*
Total revenues	wG1	*wFi2, wFi3, wFi30, wFi31*
Wastewater quality monitoring running costs	wG24	*wFi23*
Wastewater treatment running costs	wG22	*wFi21*

TIME DATA

Assessment period	wH1	*wEn3, wEn4, wEn6, wEn12, wEn13, wEn14, wEn15, wPe12, wPe17, wPe20, wPe21, wPe22, wPe23, wPe24, wPh8, wPh9, wOp1, wOp2, wOp3, wOp4, wOp5, wOp6, wOp7, wOp8, wOp9, wOp10, wOp11, wOp12, wOp13, wOp14, wOp15, wOp17, wOp18, wOp21, wOp22, wOp23, wOp24, wOp25, wOp26, wOp27, wOp28, wOp29, wOp31, wOp32, wOp33, wOp34, wOp35, wOp36, wOp37, wOp38, wOp39, wOp40, wOp41, wOp42, wOp44, wOp45, wOp46, wOp47, wOp48, wOp49, wOp50, wOp51, wOp52, wOp53, wQS10, wQS11, wQS12, wQS13, wQS14, wQS15, wQS17, wQS19, wQS20, wQS21, wQS22, wQS23, wQS24, wQS25, wQS26, wQS29, wFi1, wFi5, wFi6, wFi7, wFi8, wFi9, wFi10, wFi27, wFi32*

SECTION A - ENVIRONMENTAL DATA

wA1	**POPULATION EQUIVALENT WITH SATISFACTORY WASTEWATER TREATMENT**	Unit: ***p.e.*** Valid values: ≥ 0	**Input data** Used for indicators: *wEn1* Used for variables: none
	population equivalent that is served by *wastewater treatment plants* complying with discharge consents at the reference date. *Discharge consents refer to effluent quality standards that apply. The compliance is assessed regarding the loads and their potential environmental impacts.*		

wA2	**WASTEWATER TREATED**	Unit: m^3 Valid values: ≥ 0	**Input data or** **wA2 = wA3+wA5+wA7+wA9+wA11** Used for indicators: *wEn2, wQS5* Used for variables: none
	Wastewater treated by *wastewater treatment plants* or by *on-site system* facilities that are the responsibility of the wastewater undertaking, during the assessment period.		

wA3	**PRELIMINARY TREATED WASTEWATER**	Unit: m^3 Valid values: ≥ 0	**Input data** Used for indicators: *wEn2*(wA2), *wQS5*(wA2), *wQS6* Used for variables: wA2
	Volume of wastewater receiving only *preliminary treatment* at *wastewater treatment plants*, during the assessment period.		

wA4	**DAILY PEAK PRELIMINARY TREATED WASTEWATER**	Unit: m^3**/day** Valid values: ≥ 0	**Input data** Used for indicators: *wPh1* Used for variables: none
	Maximum daily volume of wastewater treated in *preliminary wastewater treatment plants*, during the assessment period. *This variable is the maximum of the sum of the individual daily volumes treated in the system, and not the sum of the individual daily maxima for each treatment plant.*		

wA5	**PRIMARY TREATED WASTEWATER**	Unit: m^3 Valid values: ≥ 0	**Input data** Used for indicators: *wEn2*(wA2), *wQS5*(wA2), *wQS7* Used for variables: wA2
	Volume of wastewater receiving only *primary treatment* at *wastewater treatment plants*, during the assessment period.		

wA6	**DAILY PEAK PRIMARY TREATED WASTEWATER**	Unit: m^3**/day** Valid values: ≥ 0	**Input data** Used for indicators: *wPh2* Used for variables: none
	Maximum daily volume of wastewater treated in *primary wastewater treatment plants*, during the assessment period. *This variable is the maximum of the sum of the individual daily volumes treated in the system, and not the sum of the individual daily maxima for each treatment plant.*		

wA7	**SECONDARY TREATED WASTEWATER**	Unit: m^3 Valid values: ≥ 0	**Input data** Used for indicators: *wEn2*(wA2), *wQS5*(wA2), *wQS8* Used for variables: wA2
	Volume of wastewater receiving *secondary treatment* at *wastewater treatment plants*, during the assessment period.		

wA8	**DAILY PEAK SECONDARY TREATED WASTEWATER**	Unit: ***m^3/day*** Valid values: ≥ 0	**Input data** Used for indicators: *wPh3* Used for variables: none
	Maximum daily volume of wastewater treated in *secondary wastewater treatment plants*, during the assessment period. *This variable is the maximum of the sum of the individual daily volumes treated in the system, and not the sum of the individual daily maxima for each treatment plant.*		

wA9	**TERTIARY TREATED WASTEWATER**	Unit: ***m^3*** Valid values: ≥ 0	**Input data** Used for indicators: *wEn2*(wA2), *wQS5*(wA2), *wQS9* Used for variables: wA2
	Volume of wastewater receiving *tertiary treatment* at *wastewater treatment plants*, during the assessment period.		

wA10	**DAILY PEAK TERTIARY TREATED WASTEWATER**	Unit: ***m^3/day*** Valid values: ≥ 0	**Input data** Used for indicators: *wPh4* Used for variables: none
	Maximum daily volume of wastewater treated in *tertiary wastewater treatment plants*, during the assessment period. *This variable is the maximum of the sum of the individual daily volumes treated in the system, and not the sum of the individual daily maxima for each treatment plant.*		

wA11	**WASTEWATER TREATED BY ON-SITE SYSTEMS**	Unit: ***m^3*** Valid values: ≥ 0	**Input data** Used for indicators: *wEn2*(wA2), *wQS5* Used for variables: wA2
	Volume of wastewater treated at *on-site systems*, during the assessment period.		

wA12	**WASTEWATER REUSED**	Unit: ***m^3*** Valid values: ≥ 0	**Input data** Used for indicators: *wEn2* Used for variables: none
	Treated *wastewater reused* during the assessment period.		

wA13	**SLUDGE PRODUCED IN WWTP**	Unit: ***ton DS*** Valid values: ≥ 0	**Input data** Used for indicators: *wEn6* Used for variables: none
	Dry weight of *sludge* produced in *wastewater treatment plants* managed by the undertaking during the assessment period.		

wA14	**SLUDGE HANDLED**	Unit: ***ton DS*** Valid values: ≥ 0	**Input data** Used for indicators: *wEn7*, *wEn8*, *wEn9*, *wEn10*, *wEn11* Used for variables: none
	All dry weight of *sludge* handled by the undertaking during the assessment period, including not only the dry weight of sludge produced in the *wastewater treatment plants*, but also dry weight of sludge inputs from other sources. *Sludge handled may include sludge from on-site systems.*		

wA15	**SLUDGE UTILISED**	Unit: ***ton DS*** Valid values: ≥ 0	**Input data** Used for indicators: *wEn7* Used for variables: none
	Dry weight of *sludge*, handled by the undertaking, that is utilised during the assessment period, including for example agriculture, forest, products and materials.		

wA16	**SLUDGE DISPOSED**	Unit: ***ton DS*** Valid values: ≥ 0	**Input data or** **wA16 = wA17+wA18+wA19** Used for indicators: *wEn8* Used for variables: none
	Dry weight of *sludge* handled by the undertaking that is disposed in landfill, incinerated or by any other way of disposal, during the assessment period.		

wA17	**SLUDGE GOING TO LANDFILL**	Unit: ***ton DS*** Valid values: ≥ 0	**Input data** Used for indicators: *wEn8*(wA16), *wEn9* Used for variables: wA16
	Dry weight of *sludge* going to landfill during the assessment period.		

wA18	**SLUDGE THERMALLY PROCESSED**	Unit: ***ton DS*** Valid values: ≥ 0	**Input data** Used for indicators: *wEn8*(wA16), *wEn10* Used for variables: wA16
	Dry weight of *sludge* thermally processed during the assessment period.		

wA19	**OTHER SLUDGE DISPOSAL**	Unit: ***ton DS*** Valid values: ≥ 0	**Input data** Used for indicators: *wEn8*(wA16), *wEn11* Used for variables: wA16
	Dry weight of *sludge disposed* by other means than in wA17 and wA18 during the assessment period.		

wA20	**SEWER SEDIMENTS**	Unit: ***ton*** Valid values: ≥ 0	**Input data** Used for indicators: *wEn12* Used for variables: none
	Drained weight of *sediments* removed from sewers during the assessment period.		

wA21	**ANCILLARIES SEDIMENTS**	Unit: ***ton*** Valid values: ≥ 0	**Input data** Used for indicators: *wEn13* Used for variables: none
	Drained weight of *sediments* removed from *ancillaries* during the assessment period.		

wA22	**SCREENINGS AND GRIT**	Unit: ***ton*** Valid values: ≥ 0	**Input data** Used for indicators: *wEn14* Used for variables: none
	Drained weight of screenings and grit removed in *wastewater treatment plants* and in *pumping stations* during the assessment period.		

wA23	**ON-SITE SYSTEM SEDIMENTS**	Unit: ***ton*** Valid values: ≥ 0	**Input data** Used for indicators: *wEn15* Used for variables: none
	Drained weight of *sediments* removed from *on-site systems* during the assessment period.		

wA24	**NUMBER OF OVERFLOW DISCHARGES**	Unit: ***No.*** Valid values: Integer	**Input data** Used for indicators: *wEn3* Used for variables: none
	Number of *overflow* discharges that occurred during the assessment period.		

wA25	**VOLUME OF OVERFLOW DISCHARGES**	Unit: ***m³*** Valid values: ≥ 0	**Input data**
	Total volume of *overflow* discharges that occurred during the assessment period.		Used for indicators: *wEn4, wEn5* Used for variables: none

wA26	**RAINFALL VOLUME**	Unit: ***m³*** Valid values: ≥ 0	**Input data**
	Volume of rainfall in the catchment area during the assessment period. *See section 8.3 - System profile for definition of catchment area.*		Used for indicators: *wEn5* Used for variables: none

SECTION B - PERSONNEL DATA

wB1	**TOTAL PERSONNEL**	Unit: ***No.*** Valid values: ≥ 0	**Input data**
	Total number of full time equivalent employees of the wastewater undertaking at the reference date. *Employees include every person who works for the undertaking in return for wage.*		Used for indicators: *wPe3, wPe4, wPe5, wPe6, wPe7, wPe8, wPe9, wPe14, wPe15, wPe16, wPe17, wPe20, wPe21, wPe22, wPe23, wPe24* Used for variables: wB17

wB2	**WWT PERSONNEL**	Unit: ***No.*** Valid values: ≥ 0	**Input data**
	Number of full time equivalent employees working on *wastewater treatment plants* at the reference date. *WWT refers to both end of pipe treatment plants and on-site systems that are the responsibility of the undertaking.*		Used for indicators: *wPe1* Used for variables: none

wB3	**SEWER SYSTEM PERSONNEL**	Unit: ***No.*** Valid values: ≥ 0	**Input data**
	Number of full time equivalent employees working on *sewer system* at the reference date.		Used for indicators: *wPe2* Used for variables: none

wB4	**GENERAL MANAGEMENT PERSONNEL**	Unit: ***No.*** Valid values: ≥ 0	**Input data**
	Number of full time equivalent employees dedicated to directorate, central administration, strategic planning, public relations and other *stakeholders* relations, legal affairs, internal audits, environmental management, general computing support and new business development activities, at the reference date.		Used for indicators: *wPe3* Used for variables: none

wB5	**HUMAN RESOURCES MANAGEMENT PERSONNEL**	Unit: ***No.*** Valid values: ≥ 0	**Input data**
	Number of full time equivalent employees dedicated to personnel administration, education and training, occupational safety and medicine services and social activities, at the reference date.		Used for indicators: *wPe4* Used for variables: none

wB6	FINANCIAL AND COMMERCIAL SERVICES PERSONNEL	Unit: ***No.*** Valid values: ≥ 0	**Input data** Used for indicators: *wPe5* Used for variables: none
	Number of full time equivalent employees dedicated to economic and financial planning, economic administration, economic controlling, purchasing and material management activities, at the reference date.		

wB7	CUSTOMER SERVICE PERSONNEL	Unit: ***No.*** Valid values: ≥ 0	**Input data** Used for indicators: *wPe6* Used for variables: none
	Number of full time equivalent employees dedicated to accounting and control and to customer relations and management activities, at the reference date.		

wB8	TECHNICAL SERVICES PERSONNEL	Unit: ***No.*** Valid values: ≥ 0	**Input data or** **wB8 = wB11+wB12 or** **wB8 = wB9+wB10+wB13** Used for indicators: *wPe7, wPe13* Used for variables: none
	Number of full time equivalent employees dedicated to planning, design, construction, operation, maintenance and *repair* activities, at the reference date.		

wB9	TECHNICAL WWT PERSONNEL	Unit: ***No.*** Valid values: ≥ 0	**Input data** Used for indicators: *wPe7*(wB8), *wPe10*, *wPe13*(wB8) Used for variables: wB8
	Number of full time equivalent employees working on wastewater treatment planning, design, construction, operation and maintenance, at the reference date. *In this PI, wastewater treatment refers to both end of pipe wastewater treatment plants and on-site systems that are the responsibility of the undertaking. Personnel dealing with quality sampling and testing included in the undertaking laboratories activities shall not be accounted for in this variable. See definitions of section 4.5 - Organisational functions and definitions.*		

wB10	TECHNICAL SE PERSONNEL	Unit: ***No.*** Valid values: ≥ 0	**Input data** Used for indicators: *wPe7*(wB8), *wPe11*, *wPe13*(wB8) Used for variables: wB8
	Number of full time equivalent employees working on *sewer system* planning, design, operation and maintenance, at the reference date. *Personnel dealing with quality sampling and testing included in the undertaking laboratories activities shall not be accounted for in this variable. See definitions of section 4.5 - Organisational functions and definitions.*		

wB11	PLANNING, DESIGN AND CONSTRUCTION PERSONNEL	Unit: ***No.*** Valid values: ≥ 0	**Input data** Used for indicators: *wPe7*(wB8), *wPe8*, *wPe13*(wB8) Used for variables: wB8
	Number of full time equivalent employees dedicated to planning, design and construction of *sewer system* and wastewater treatment facilities, at the reference date. *See definitions of section 4.5 - Organisational functions and definitions.*		

wB12	OPERATION AND MAINTENANCE PERSONNEL	Unit: ***No.*** Valid values: ≥ 0	**Input data** Used for indicators: *wPe7*(wB8), *wPe9*, *wPe13*(wB8), *wPe18*, *wPe19*, *wOp55* Used for variables: wB8
	Number of full time equivalent employees dedicated to operation, maintenance and *repair* of *sewer system* and wastewater treatment facilities, at the reference date.		

wB13	**LABORATORY PERSONNEL**	Unit: ***No.*** Valid values: ≥ 0	**Input data**
	Total number of full time equivalent employees working on wastewater quality sampling and testing, at the reference date. *In the case of multi-utilities, each employee shall be accounted for based on the proportion of time dedicated to wastewater related activities.* *See definitions of section 4.5 - Organisational functions and definitions.* *For the ease of assessment, this variable can be broken down into more detailed sub-variables.* *If laboratory testing is totally outsourced, employees working on wastewater sampling shall be accounted for as WWT or SE personnel.*		Used for indicators: *wPe7*(wB8), *wPe12*, *wPe13*(wB8) Used for variables: wB8

wB14	**SUPPORT SERVICES PERSONNEL**	Unit: ***No.*** Valid values: ≥ 0	**Input data**
	Total number of full time equivalent employees working on the central stock repository, central workshops and the vehicle fleet, at the reference date. *In the case of multi-utilities, each employee shall be accounted for based on the proportion of time dedicated to wastewater related activities.* *See definitions of section 4.5 - Organisational functions and definitions.* *For the ease of assessment, this variable can be broken down into more detailed sub-variables.*		Used for indicators: *wPe13* Used for variables: none

wB15	**UNIVERSITY DEGREE PERSONNEL**	Unit: ***No.*** Valid values: ≥ 0	**Input data**
	Total number of full time equivalent employees with university degree, at the reference date. *In the case of multi-utilities, each employee shall be accounted for based on the proportion of time dedicated to wastewater related activities. Employees may include permanent and temporary personnel.*		Used for indicators: *wPe14*, *wPe16*(wB17) Used for variables: wB17

wB16	**BASIC EDUCATION PERSONNEL**	Unit: ***No.*** Valid values: ≥ 0	**Input data**
	Total number of full-time equivalent employees without university degree but with basic education at least, at the reference date. *The definition of basic education to be used shall refer to the country requirements at the reference date. Definitions may vary from country to country. In the case of multi-utilities, each employee shall be accounted for based on the proportion of time dedicated to wastewater related activities. Employees may include permanent and temporary personnel.*		Used for indicators: *wPe15*, *wPe16*(wB17) Used for variables: wB17

wB17	**OTHER QUALIFICATION PERSONNEL**	Unit: ***No.*** Valid values: ≥ 0	**Input data or** **wB17 = wB1 - (wB15+wB16)**
	Total number of full-time equivalent employees without basic education, at the reference date or Total personnel - (university degree personnel + basic education personnel). *The definition of basic education to be used shall refer to the country requirements at the reference date. Definitions may vary from country to country. In the case of multi-utilities, each employee shall be accounted for based on the proportion of time dedicated to wastewater related activities. Employees may include permanent and temporary personnel.*		Used for indicators: *wPe16* Used for variables: none

wB18	**CONFINED SPACES SKILLED PERSONNEL**	Unit: ***No.*** Valid values: ≥ 0	**Input data** Used for indicators: *wPe19* Used for variables: none
	Number of full time equivalent wastewater undertaking employees with appropriate authorised confined spaces training, at the reference date. *"Authorised confined spaces training" is recommended for all staff who are required to enter a designated confined space for purposes of inspection, cleaning, maintenance, repair or any form of work. Working in confined spaces can lead to injury or even death if adequate precautions are not taken. Therefore only trained persons should enter or work in confined spaces.* *A confined space is a space with restricted ventilation where special safety precautions may need to be taken because the atmosphere is liable to be contaminated at any time (by dust, fumes, mist, vapour, gas or other harmful substance) or is liable to be oxygen deficient (EN 752-1, 1995 and ASOTT, 2002).*		

wB19	**VACCINATED PERSONNEL**	Unit: ***No.*** Valid values: ≥ 0	**Input data** Used for indicators: *wPe18* Used for variables: none
	Number of full time equivalent employees with up to date vaccination certification for *sewage* related diseases, at the reference date.		

wB20	**TRAINING TIME**	Unit: ***Hour*** Valid values: ≥ 0	**Input data** Used for indicators: *wPe17* Used for variables: none
	Number of training hours for wastewater personnel, during the assessment period. *In the case of multi-utilities, a proportion of the wastewater in the overall productive activities has to be established. For the accounting of this variable, the training hours provided to employees working in general activities shall be affected by this proportion.* *Employees may include permanent and temporary personnel.*		

wB21	**WORKING ACCIDENTS**	Unit: ***No.*** Valid values: Integer	**Input data** Used for indicators: *wPe20* Used for variables: none
	Number of working accidents requiring medical care occurring with personnel, during the assessment period. *If the wastewater undertaking is responsible for other productive activities beyond wastewater, a proportion of the wastewater in the overall productive activities has to be established.*		

wB22	**WORKING FATALITIES**	Unit: ***No.*** Valid values: Integer	**Input data** Used for indicators: *wPe21* Used for variables: none
	Number of working accidents resulting on death or permanent disease, during the assessment period.		

wB23	**ABSENTEEISM**	Unit: ***Day*** Valid values: ≥ 0	**Input data** Used for indicators: *wPe22*, *wPe24*(wB25) Used for variables: wB25
	Number of days of personnel absenteeism, during the assessment period.		

wB24	**ABSENTEEISM DUE TO ACCIDENTS OR ILLNESS AT WORK**	Unit: ***Day*** Valid values: ≥ 0	**Input data**
	Number of days of personnel absenteeism due to accidents or illness, during the assessment period.		Used for indicators: *wPe23*, *wPe24*(wB25) Used for variables: wB25

wB25	**ABSENTEEISM DUE TO OTHER REASONS**	Unit: ***Day*** Valid values: ≥ 0	**Input data or** **wB25 = wB23 - wB24**
	Number of days of personnel absenteeism that were not due to working accidents or illness, during the assessment period.		Used for indicators: *wPe24* Used for variables: none

wB26	**NORMAL WORK**	Unit: ***Hour*** Valid values: ≥ 0	**Input data**
	Number of working hours of all the undertaking employees, during the assessment period (employed hours gross - absence for holidays).		Used for indicators: *wPe25* Used for variables: none

wB27	**OVERTIME WORK**	Unit: ***Hour*** Valid values: ≥ 0	**Input data**
	Total number of overtime working hours of the undertaking employees, during the assessment period.		Used for indicators: *wPe25* Used for variables: none

SECTION C - PHYSICAL ASSETS DATA

wC1	**TOTAL SEWER LENGTH**	Unit: ***km*** Valid values: ≥ 0	**Input data**
	Total length of *sewers* managed by the undertaking at the reference date. *service connections* excluded.		Used for indicators: *wEn12*, *wEn13*, *wEn14*, *wPe2*, *wPe11*, *wPh5*, *wPh6*, *wPh7*, *wOp1*, *wOp2*, *wOp21*, *wOp22*, *wOp23*, *wOp24*, *wOp31*, *wOp32*, *wOp33*, *wOp34*, *wOp35*, *wOp37*, *wOp38*, *wOp39*, *wOp40*, *wOp54*, *wFi6*, *wFi8*, *wFi10* Used for variables: none

wC2	**SURCHARGED SEWERS IN DRY WEATHER**	Unit: ***m*** Valid values: ≥ 0	**Input data**
	Length of *sewer* where *surcharging* has occurred in *dry weather*, during the assessment period. *service connections* excluded.		Used for indicators: *wPh5* Used for variables: none

wC3	**SURCHARGED SEWERS IN WET WEATHER**	Unit: ***m*** Valid values: ≥ 0	**Input data**
	Length of *sewer* where *surcharging* has occurred in *wet weather*, during the assessment period. *service connections* excluded.		Used for indicators: *wPh6* Used for variables: none

wC4	HIGHLY SURCHARGED SEWERS	Unit: ***m*** Valid values: ≥ 0	**Input data**
	Length of *sewer* where high degree of *surcharging* has occurred. *High degree surcharging means water at least 0.5 m above the pipe crown (pipe crown is the top of the sewer / external; the top of the sewer inside the pipe is termed 'soffit').* *service connections* excluded.		Used for indicators: *wPh7* Used for variables: none

wC5	WWTP PUMPS POWER	Unit: ***kW*** Valid values: ≥ 0	**Input data**
	Total nominal power of the *wastewater treatment plant* pumps at the reference date. *Wastewater treatment plant inlet pumps are to be included. Pump power installed does not include standby pumps (those used for substitution).*		Used for indicators: *wPh9* Used for variables: none

wC6	WASTEWATER SYSTEM PUMPING STATIONS	Unit: ***No.*** Valid values: Integer	**Input data**
	Number of wastewater *pumping stations* in the *sewer system*, including *wastewater treatment plant* inlet pumping stations, at the reference date.		Used for indicators: *wOp10, wOp36, wOp42* Used for variables: none

wC7	WASTEWATER SYSTEM PUMPS	Unit: ***No.*** Valid values: Integer	**Input data**
	Total number of pumps in the wastewater system managed by the undertaking, at the reference date. *When accounting for the total number of pumps of the system, small pumps can be excluded if their influence in terms of the system reliability is negligible.*		Used for indicators: *wOp41* Used for variables: none

wC8	WASTEWATER SYSTEM PUMPS POWER	Unit: ***kW*** Valid values: ≥ 0	**Input data**
	Total nominal power of the wastewater pumps in the *sewer system*, including *wastewater treatment plant* inlet *pumping stations*, at the reference date. *Pump power installed does not include standby pumps (those used for substitution).*		Used for indicators: *wOp28, wOp29* Used for variables: none

wC9	SEWER SYSTEM PUMPING STATIONS	Unit: ***No.*** Valid values: Integer	**Input data**
	Total number of *sewer system pumping stations* at the reference date. *This variable does not include pumping stations in wastewater treatment plants.*		Used for indicators: *wPh10* Used for variables: none

wC10	SEWER SYSTEM PUMPS POWER	Unit: ***kW*** Valid values: ≥ 0	**Input data**
	Total nominal power of the *sewer system* pumps at the reference date. *Pump power installed does not include standby pumps (those used for substitution).*		Used for indicators: *wPh8, wOp11* Used for variables: none

wC11	**SEWER SYSTEM PUMPING STATION UTILISATION**	Unit: ***No.*** Valid values: Integer	**Input data** Used for indicators: *wPh10* Used for variables: none
	Number of *pumping stations* in the *sewer systems* where pumps operated more than 75% of the time during the assessment period. *The percentage of operating time of each pumping station is assessed by dividing the total number of operating hours of the installed pumps during the assessment period by the maximum annual operating hours for the installed pumps (number of pumps x 24 x wH1).*		

wC12	**DAILY PRELIMINARY TREATMENT CAPACITY**	Unit: m^3***/day*** Valid values: ≥ 0	**Input data** Used for indicators: *wPh1* Used for variables: none
	Sum, for all *preliminary wastewater treatment plants*, of the daily maximum treatment capacities (evaluated during the assessment period).		

wC13	**DAILY PRIMARY TREATMENT CAPACITY**	Unit: m^3***/day*** Valid values: ≥ 0	**Input data** Used for indicators: *wPh2* Used for variables: none
	Sum, for all *primary wastewater treatment plants*, of the daily maximum treatment capacities (evaluated during the assessment period).		

wC14	**DAILY SECONDARY TREATMENT CAPACITY**	Unit: m^3***/day*** Valid values: ≥ 0	**Input data** Used for indicators: *wPh3* Used for variables: none
	Sum, for all *secondary wastewater treatment plants*, of the daily maximum treatment capacities (evaluated during the assessment period).		

wC15	**DAILY TERTIARY TREATMENT CAPACITY**	Unit: m^3***/day*** Valid values: ≥ 0	**Input data** Used for indicators: *wPh4* Used for variables: none
	Sum, for all *tertiary wastewater treatment plants*, of the daily maximum treatment capacities (evaluated during the assessment period).		

wC16	**CONTROL UNITS**	Unit: ***No.*** Valid values: Integer	**Input data** Used for indicators: *wPh11, wPh12* Used for variables: none
	Total number of control units, at the reference date, defined as functionally independent sets of control devices of the system (e.g. control valves or pump switches) used for wastewater collection, pumping, *storage*, treatment and energy recovery. *For example, the set of valves and other control devices used for controlling multiple pumping and storage installations shall be considered a single control unit. A similar criterium can be used, for instance, for each unit process in treatment plants and associated pumping and storage facilities (e.g. trickling filters).*		

wC17	**AUTOMATED CONTROL UNITS**	Unit: ***No.*** Valid values: Integer	**Input data** Used for indicators: *wPh11* Used for variables: none
	Total number of automated control units of the system, at the reference date, defined as functionally independent sets of control devices of the system (e.g. control valves or pump switches) that are locally controlled by a programmable logical controller, used for wastewater collection, pumping, *storage*, treatment and energy recovery.		

wC18	**REMOTELY CONTROLLED UNITS**	Unit: ***No.*** Valid values: Integer	**Input data** Used for indicators: *wPh12* Used for variables: none
	Total number of remote control units of the wastewater system connected to a central control system, at the reference date, usually via information cabling or radio connection.		

wC19	**OVERFLOW DEVICES**	Unit: ***No.*** Valid values: Integer	**Input data** Used for indicators: *wEn3, wEn4* Used for variables: none
	Number of *overflow* devices in the *sewer system*, at the reference date.		

wC20	**NUMBER OF CSOs**	Unit: ***No.*** Valid values: Integer	**Input data** Used for indicators: *wOp6, wOp43* Used for variables: none
	Number of *combined sewer overflow (CSO)* devices in the *sewer system* at the reference date.		

wC21	**MANHOLE CHAMBERS**	Unit: ***No.*** Valid values: Integer	**Input data** Used for indicators: *wOp3, wOp25, wOp26* Used for variables: none
	Number of manhole chambers in the *sewer system* at the reference date.		

wC22	**GULLY POTS**	Unit: ***No.*** Valid values: Integer	**Input data** Used for indicators: *wOp4, wOp5* Used for variables: none
	Number of *gully pots* in the *sewer system* at the reference date.		

wC23	**NUMBER OF STORAGE TANKS**	Unit: ***No.*** Valid values: Integer	**Input data** Used for indicators: *wOp6* Used for variables: none
	Number of *storage tanks* (in or off line) in the *sewer system* at the reference date.		

wC24	**VOLUME OF TANKS AND CSOs**	Unit: m^3 Valid values: ≥ 0	**Input data** Used for indicators: *wOp7, wOp8* Used for variables: none
	Total volume of *storage tanks* and *CSOs* at the reference date.		

wC25	**SCREENS OF TANKS AND CSOs**	Unit: ***No.*** Valid values: Integer	**Input data** Used for indicators: *wOp9* Used for variables: none
	Number of *screens* associated with tanks and *CSOs* at the reference date.		

wC26	**WWTP FLOWMETERS**	Unit: ***No.*** Valid values: Integer	**Input data** Used for indicators: *wOp13* Used for variables: none
	Number of permanently installed flowmeters in *wastewater treatment plants* at the reference date.		

wC27	**SE FLOWMETERS**	Unit: ***No.*** Valid values: Integer	**Input data** Used for indicators: *wOp12* Used for variables: none
	Number of permanently installed flowmeters in the *sewer system* at the reference date.		

wC28	**CONNECTED PROPERTIES**	Unit: ***No.*** Valid values: Integer	**Input data** Used for indicators: *wQS10, wQS11, wQS12, wQS13, wQS14, wQS15* Used for variables: none
	Number of properties connected to the *sewer system* managed by the undertaking, at the reference date.		

wC29	**SERVICE CONNECTIONS**	Unit: ***No.*** Valid values: Integer	**Input data** Used for indicators: *wOp27* Used for variables: none
	Total number of *service connections* at the reference date.		

wC30	**EMERGENCY POWER SYSTEMS**	Unit: ***kW*** Valid values: ≥ 0	**Input data** Used for indicators: *wOp15* Used for variables: none
	Total nominal power of the emergency power systems at the reference date.		

wC31	**SIGNAL TRANSMISSION EQUIPMENT**	Unit: ***No.*** Valid values: Integer	**Input data** Used for indicators: *wOp16* Used for variables: none
	Total number of signal transmission equipments at the reference date.		

wC32	**ELECTRICAL SWITCHGEAR**	Unit: ***No.*** Valid values: Integer	**Input data** Used for indicators: *wOp17* Used for variables: none
	Total number of electrical switchgear equipments at the reference date.		

wC33	**WASTEWATER QUALITY MONITORING INSTRUMENTS**	Unit: ***No.*** Valid values: Integer	**Input data** Used for indicators: *wOp14* Used for variables: none
	Number of wastewater quality instruments installed permanently in *wastewater treatment plants*, at the reference date.		

SECTION D - OPERATIONAL DATA

wD1	**SEWER INSPECTION**	Unit: ***km*** Valid values: ≥ 0	**Input data** Used for indicators: *wOp1* Used for variables: none
	Length of *sewers inspected* during the assessment period. *If a pipe is inspected more than once in the assessment period, it shall be accounted only once. This convention may cause bias to the indicator when a non-annual assessment period is adopted. On-line inspection by remote surveillance and alarms and TV inspection shall be included.*		

wD2	**SEWER CLEANING**	Unit: ***km*** Valid values: ≥ 0	**Input data** Used for indicators: *wOp2* Used for variables: none
	Length of *sewers cleaned* for prevention of clogging and *blockage* during the assessment period. *Sewer cleaning refers to actions under proactive management strategy. If a pipe is inspected more than once in the assessment period, it shall be accounted only once. This convention may cause bias to the indicator when a non-annual assessment period is adopted. Root cutting shall be included. Curative cleaning due to blockage is reflected in wOp34.*		

wD3	**MANHOLE CHAMBER INSPECTION**	Unit: ***No.*** Valid values: Integer	**Input data** Used for indicators: *wOp3* Used for variables: none
	Number of manhole chambers *inspected* during the assessment period. *If a manhole chamber is inspected more than once in the assessment period, it shall be accounted as many times as the number of inspections.*		

wD4	**GULLY POTS INSPECTION**	Unit: ***No.*** Valid values: Integer	**Input data** Used for indicators: *wOp4* Used for variables: none
	Number of *gully pots inspected* during the assessment period. *Gully pots inspection refers to actions under proactive management strategy. If a gully pot is inspected more than once in the assessment period, it shall be accounted as many times as the number of inspections.*		

wD5	**GULLY POTS CLEANING**	Unit: ***No.*** Valid values: Integer	**Input data** Used for indicators: *wOp5* Used for variables: none
	Number of *gully pots cleaned* during the assessment period. *Gully pots clean refer to actions under proactive management strategy. If a gully pot is cleaned more than once in the assessment period, it shall be accounted as many times as the number of cleanings.*		

wD6	**STORAGE TANKS AND CSOs INSPECTION NUMBER**	Unit: ***No.*** Valid values: Integer	**Input data** Used for indicators: *wOp6* Used for variables: none
	Number of *inspections* to *storage tanks* and *CSOs* carried out during the assessment period. *If a tank or CSO is inspected more than once in the assessment period, it shall be accounted for as many times as the number of inspections.*		

wD7	**TANKS AND CSOs INSPECTED VOLUME**	Unit: ***m^3*** Valid values: ≥ 0	**Input data** Used for indicators: *wOp7* Used for variables: none
	Volume of tanks and *CSOs inspected* during the assessment period. *If a tank or CSO is inspected more than once in the assessment period, it shall be accounted for as many times as the number of inspections.*		

wD8	**TANKS AND CSOs CLEANING**	Unit: ***m^3*** Valid values: ≥ 0	**Input data** Used for indicators: *wOp8* Used for variables: none
	Volume of tanks and *CSOs cleaned* during the assessment period. *If a tank or CSO is cleaned more than once in the assessment period, it shall be accounted for as many times as the number of times cleaned. Only CSOs with storage capacity are included.*		

wD9	SCREEN INSPECTION IN TANKS AND CSOs	Unit: ***No.*** Valid values: Integer	**Input data** Used for indicators: *wOp9* Used for variables: none
	Number of *screens* associated with tanks and *CSOs inspected* during the assessment period. *If a screen is inspected more than once in the assessment period, it shall be accounted as many times as the number of inspections.*		

wD10	PUMPING STATION INSPECTION	Unit: ***No.*** Valid values: Integer	**Input data** Used for indicators: *wOp10* Used for variables: none
	Number of *inspections* to *pumping stations* carried out during the assessment period. *If a pump is inspected more than once in a year, it shall be accounted as many times as the number of inspections.*		

wD11	PUMP INSPECTION	Unit: ***kW*** Valid values: ≥ 0	**Input data** Used for indicators: *wOp11* Used for variables: none
	Sum of the nominal power of pumps and related *ancillaries* subject to *inspection* during the assessment period. *Pumps integrated in the treatment process are excluded. If a pump is inspected more than once in a year, it shall be accounted as many times as the number of inspections.*		

wD12	PUMP ENERGY CONSUMPTION	Unit: ***kWh*** Valid values: ≥ 0	**Input data or** **wD12 = wD14 + wD15** Used for indicators: *wOp20* Used for variables: none
	$\Sigma_{\text{for all pumps installed}}$(pump nominal power x pump working hours during the assessment period).		

wD13	WWT ENERGY CONSUMPTION	Unit: ***kWh*** Valid values: ≥ 0	**Input data** Used for indicators: *wOp18*, *wOp19* Used for variables: none
	Energy consumed during the assessment period by wastewater treatment processes managed by the undertaking. *WWT includes on-site systems; thermal processing of sludge should not be included.*		

wD14	PUMP ENERGY CONSUMPTION IN WWTP	Unit: ***kWh*** Valid values: ≥ 0	**Input data** Used for indicators: *wPh9*, *wOp20*(wD12) Used for variables: wD12
	$\Sigma_{\text{for all pumps installed in WWTP}}$(pump nominal power x pump working hours during the assessment period). *wastewater treatment plant inlet pumps are to be included.*		

wD15	PUMP ENERGY CONSUMPTION IN SE	Unit: ***kWh*** Valid values: ≥ 0	**Input data** Used for indicators: *wPh8*, *wOp20*(wD12) Used for variables: wD12
	$\Sigma_{\text{for all pumps installed in SE}}$(pump nominal power x pump working hours during the assessment period).		

wD16	STANDARDISATION FACTOR	Unit: ***m^3 x m*** Valid values: Real	**Input data** Used for indicators: *wOp20* Used for variables: none
	Sum, for all the pumps of the system, of D2(i), D2(i) being: D2(i) = V(i) x h(i), where V is the total volume (m^3) pumped by pump i during the assessment period and h(i) is the pump head (m).		

wD17	CO-GENERATED ENERGY	Unit: ***kWh*** Valid values: ≥ 0	**Input data** Used for indicators: *wOp19* Used for variables: none
	Energy produced by co-generation processes during the assessment period.		

wD18	POWER FAILURES	Unit: ***hour*** Valid values: ≥ 0	**Input data** Used for indicators: *wOp42* Used for variables: none
	Sum, for all *pumping stations*, of the number of hours each pumping station was out of service during the assessment period, due to power supply interruptions. *This variable assesses the time that pumping stations are out of service due to power failures, regardless of the power source used (normal or standby power generation system). This means that the time when a pumping station is reliant on standby power generation shall not be considered as a power failure. This indicator aims to measure the performance of the wastewater undertaking and not of the power supply undertaking.*		

wD19	WWTP FLOW METER CALIBRATIONS	Unit: ***No.*** Valid values: Integer	**Input data** Used for indicators: *wOp13* Used for variables: none
	Number of *calibrations* carried out in flow meters permanently installed in *wastewater treatment plants* during the assessment period. *If a flowmeter is calibrated more than once in a year, it shall be accounted for as many times as the number of calibrations.*		

wD20	SE FLOW METER CALIBRATIONS	Unit: ***No.*** Valid values: Integer	**Input data** Used for indicators: *wOp12* Used for variables: none
	Number of *calibrations* carried out in flow meters permanently installed in the *sewer system* during the assessment period. *If a flowmeter is calibrated more than once in a year, it shall be accounted for as many times as the number of calibrations.*		

wD21	WASTEWATER QUALITY MONITORING INSTRUMENTS CALIBRATIONS	Unit: ***No.*** Valid values: Integer	**Input data** Used for indicators: *wOp14* Used for variables: none
	Number of permanent automatic wastewater quality monitoring instrument *calibrations* carried out in *wastewater treatment plants* during the assessment period. *If an equipment is calibrated more than once in a year, it shall be accounted for as many times as the number of calibrations.*		

wD22	**EMERGENCY POWER SYSTEMS UNITS INSPECTION**	Unit: ***kW*** Valid values: ≥ 0	**Input data** Used for indicators: *wOp15* Used for variables: none
	Sum of the nominal power of the system emergency power equipment *inspected* during the assessment period. *If an equipment is inspected more than once during the assessment period, it shall be accounted for as many times as the number of inspections.*		

wD23	**SIGNAL TRANSMISSION UNITS INSPECTION**	Unit: ***No.*** Valid values: Integer	**Input data** Used for indicators: *wOp16* Used for variables: none
	Number of signal transmission units *inspected* during the assessment period. *If an equipment is inspected more than once during the assessment period, it shall be accounted for as many times as the number of inspections.*		

wD24	**ELECTRICAL SWITCHGEAR INSPECTION**	Unit: ***No.*** Valid values: Integer	**Input data** Used for indicators: *wOp17* Used for variables: none
	Number of electrical switchgear units *inspected* during the assessment period. *If an equipment is inspected more than once during the assessment period, it shall be accounted for as many times as the number of inspections.*		

wD25	**SEWER REHABILITATION**	Unit: ***km*** Valid values: ≥ 0	**Input data** Used for indicators: *wOp21* Used for variables: none
	Length of *sewers rehabilitated* or *renewed* during the assessment period. *This variable includes not only wD26 and wD27 but also the length of sewers rehabilitated with other techniques. For the assessment of this variable, only the length of defect shall be considered.*		

wD26	**SEWER RENOVATION**	Unit: ***km*** Valid values: ≥ 0	**Input data** Used for indicators: *wOp22* Used for variables: none
	Length of *sewers renovated* (e.g. *relined*) during the assessment period. *service connections excluded.*		

wD27	**SEWER REPLACEMENT**	Unit: ***km*** Valid values: ≥ 0	**Input data** Used for indicators: *wOp23* Used for variables: none
	sewer length *replaced* (or *renewed*) during the assessment period. *service connections excluded.*		

wD28	**SEWER AND JOINTS REPAIR**	Unit: ***No.*** Valid values: ≥ 0	**Input data** Used for indicators: *wOp24* Used for variables: none
	Number of *repairs* carried out in defective *sewers* and joints during the assessment period. *service connections excluded. Connection joints included.*		

wD29	**MANHOLE CHAMBER REPLACEMENT, RENEWAL, RENOVATION OR REPAIR**	Unit: ***No.*** Valid values: Integer	**Input data** Used for indicators: *wOp25* Used for variables: none
	Number of manhole chambers *replaced*, *renewed*, *renovated* or *repaired* during the assessment period.		

wD30	**MANHOLE COVERS REPLACEMENT**	Unit: ***No.*** Valid values: Integer	**Input data** Used for indicators: *wOp26* Used for variables: none
	Number of manhole covers *replaced* during the assessment period.		

wD31	**SERVICE CONNECTION REPLACEMENT OR RENEWAL**	Unit: ***No.*** Valid values: Integer	**Input data** Used for indicators: *wOp27* Used for variables: none
	Number of *service connections replaced* or *renewed* during the assessment period.		

wD32	**PUMP REFURBISMENT**	Unit: ***kW*** Valid values: ≥ 0	**Input data** Used for indicators: *wOp28* Used for variables: none
	Total nominal power of pumps subject to overhaul (i.e. *refurbishment* or *renewal* of relevant elements necessary to bring unit back to original performance), including *wastewater treatment plant* inlet *pumping stations*, during the assessment period. *Refurbishment may include partial replacement.*		

wD33	**PUMP REPLACEMENT**	Unit: ***kW*** Valid values: ≥ 0	**Input data** Used for indicators: *wOp29* Used for variables: none
	Total nominal power of pumps *replaced*, including *wastewater treatment plant* inlet *pumping stations*, during the assessment period. *This variable refers exclusively to full replacement of the pump and the engine. The nominal power of the new equipment shall be accounted for.*		

wD34	**PUMPS OUT OF ORDER**	Unit: ***Hour*** Valid values: ≥ 0	**Input data** Used for indicators: *wOp41* Used for variables: none
	Sum, for all pumps, of the number of hours when each pump was out of order, during the assessment period. *When accounting for the total number of pumps of the system, small pumps can be excluded if their influence in terms of the system reliability is negligible.*		

wD35	*inflow* **VOLUME**	Unit: ***m^3*** Valid values: ≥ 0	**Input data** Used for indicators: *wOp30, wOp31* Used for variables: none
	Volume of water entering *sewers* from *wrong connections* during the assessment period. *Wrong connections include incorrect connections to the sewer system (e.g. direct surface runoff and entry through manhole covers to sanitary sewers).*		

wD36	*infiltration* **VOLUME**	Unit: ***m³*** Valid values: ≥ 0	**Input data** Used for indicators: *wOp30, wOp32* Used for variables: none
	Volume of water entering *sewers* from groundwater during the assessment period.		

wD37	*exfiltration* **VOLUME**	Unit: ***m³*** Valid values: ≥ 0	**Input data** Used for indicators: *wOp30, wOp33* Used for variables: none
	Volume of leakage from *sewers* into the ground during the assessment period.		

wD38	**SEWER BLOCKAGES**	Unit: ***No.*** Valid values: Integer	**Input data** Used for indicators: *wOp34* Used for variables: none
	Number of *blockages* that occurred in *sewers* during the assessment period. *pumping station blockages shall not be included. Include blockages in service connections only where these are the responsibility of the wastewater undertaking.*		

wD39	**SEWER BLOCKAGE LOCATIONS**	Unit: ***No.*** Valid values: Integer	**Input data** Used for indicators: *wOp35* Used for variables: none
	Number of individual locations in *sewers* where *blockages* occurred during the assessment period. *Locations where frequent blockages occurred shall only be accounted once; pumping station blockages shall not be included. Include blockage locations in service connections only where these are the responsibility of the wastewater undertaking.*		

wD40	**PUMPING STATION BLOCKAGES**	Unit: ***No.*** Valid values: Integer	**Input data** Used for indicators: *wOp36* Used for variables: none
	Number of *blockages* that occurred in *pumping stations* during the assessment period. *Pumping station blockages include blockages in pumps and valves. Sewer blockages shall not be included. If blockages occur more than once in a pumping station, this shall be accounted for as many times as the number of blockages.*		

wD41	**FLOODINGS FROM SANITARY SEWERS**	Unit: ***No.*** Valid values: Integer	**Input data** Used for indicators: *wOp37* Used for variables: none
	Number of *sanitary floodings* that occurred during the assessment period. *Include only incidents related to sanitary sewers that are the responsibility of the wastewater undertaking.*		

wD42	**FLOODINGS FROM COMBINED SEWERS**	Unit: ***No.*** Valid values: Integer	**Input data** Used for indicators: *wOp38* Used for variables: none
	Number of *floodings* that occurred from combined *sewers* during the assessment period. *Include only incidents related to combined sewers that are the responsibility of the wastewater undertaking.*		

wD43	**SURFACE FLOODINGS**	Unit: ***No.*** Valid values: Integer	**Input data** Used for indicators: *wOp39* Used for variables: none
	Number of *surface floodings* that occurred during the assessment period. *These include only surface flooding due to inadequacy of storm drainage system (combined sewers included) that is the responsibility of the wastewater undertaking. Inadequacy relates to all causes (e.g. design, operation, etc.).*		

wD44	**SEWER COLLAPSES**	Unit: ***No.*** Valid values: Integer	**Input data** Used for indicators: *wOp40* Used for variables: none
	Number of *sewer collapses* that occurred during the assessment period. *Does not include collapses on service connections.*		

wD45	**CSOs CONTROLLED**	Unit: ***No.*** Valid values: Integer	**Input data** Used for indicators: *wOp43* Used for variables: none
	Number of *CSOs* controlled by dynamic or static *flow regulators* at the reference date.		

wD46	**WASTEWATER QUALITY TESTS CARRIED OUT**	Unit: ***No.*** Valid values: Integer	**Input data or** **wD46 = wD48+wD49+wD50+wD51+ wD52+wD53+wD54** Used for indicators: *wOp44* Used for variables: none
	Total number of wastewater quality tests carried out during the assessment period. *This variable shall only account for tests that are the responsibility of the undertaking, at points where quality monitoring is mandatory, wherever they are carried out. Includes not only tests performed in the undertaking laboratories, but also the outsourced ones.*		

wD47	**WASTEWATER QUALITY TESTS CARRIED OUT BY UNDERTAKING LABORATORIES**	Unit: ***No.*** Valid values: Integer	**Input data** Used for indicators: *wPe12* Used for variables: none
	Total number of wastewater and *sludge* quality tests carried out by the undertaking laboratories, during the assessment period. *This variable includes not only the mandatory wastewater and sludge tests but also process control tests. Outsourced tests shall not be included.*		

wD48	**BOD TESTS CARRIED OUT**	Unit: ***No.*** Valid values: Integer	**Input data** Used for indicators: *wOp44*(wD46), *wOp45* Used for variables: wD46
	Number of BOD tests carried out during the assessment period. *This variable shall only account for tests that are the responsibility of the undertaking, at points where quality monitoring is mandatory. Includes not only tests performed in the undertaking laboratories, but also the outsourced ones.*		

wD49	**COD TESTS CARRIED OUT**	Unit: ***No.*** Valid values: Integer	**Input data** Used for indicators: *wOp44*(wD46), *wOp46* Used for variables: wD46
	Number of COD tests carried out during the assessment period. *This variable shall only account for tests that are the responsibility of the undertaking, at points where quality monitoring is mandatory. Includes not only tests performed in the undertaking laboratories, but also the outsourced ones.*		

wD50	**TSS TESTS CARRIED OUT**	Unit: ***No.*** Valid values: Integer	**Input data** Used for indicators: *wOp44*(wD46), *wOp47* Used for variables: wD46
	Number of TSS tests carried out during the assessment period. *This variable shall only account for tests that are the responsibility of the undertaking, at points where quality monitoring is mandatory. Includes not only tests performed in the undertaking laboratories, but also the outsourced ones.*		

wD51	**TOTAL PHOSPHORUS TESTS CARRIED OUT**	Unit: ***No.*** Valid values: Integer	**Input data** Used for indicators: *wOp44*(wD46), *wOp48* Used for variables: wD46
	Number of phosphorus tests carried out during the assessment period. *This variable shall only account for tests that are the responsibility of the undertaking, at points where quality monitoring is mandatory. Includes not only tests performed in the undertaking laboratories, but also the outsourced ones.*		

wD52	**TOTAL NITROGEN TESTS CARRIED OUT**	Unit: ***No.*** Valid values: Integer	**Input data** Used for indicators: *wOp44*(wD46), *wOp49* Used for variables: wD46
	Number of nitrogen tests carried out by the undertaking laboratories, assessing nitrogen and nitrogen compounds, during the assessment period. *This variable shall only account for tests that are the responsibility of the undertaking, at points where quality monitoring is mandatory. Includes not only tests performed in the undertaking laboratories, but also the outsourced ones.*		

wD53	**FAECAL *E.COLI* TESTS CARRIED OUT**	Unit: ***No.*** Valid values: Integer	**Input data** Used for indicators: *wOp44*(wD46), *wOp50* Used for variables: wD46
	Number of faecal *Escherichia coli* tests carried out during the assessment period. *This variable shall only account for tests that are the responsibility of the undertaking, at points where quality monitoring is mandatory. Includes not only tests performed in the undertaking laboratories, but also the outsourced ones.*		

wD54	**OTHER WASTEWATER QUALITY TESTS CARRIED OUT**	Unit: ***No.*** Valid values: Integer	**Input data** Used for indicators: *wOp44*(wD46), *wOp51* Used for variables: wD46
	Number of other wastewater quality tests carried out during the assessment period. *This variable shall only account for tests that are the responsibility of the undertaking, at points where quality monitoring is mandatory. Includes not only tests performed in the undertaking laboratories, but also the outsourced ones.*		

wD55	**SLUDGE TESTS CARRIED OUT**	Unit: ***No.*** Valid values: Integer	**Input data** Used for indicators: *wOp52* Used for variables: none
	Number of *sludge* tests carried out during the assessment period. *Includes not only tests performed in the undertaking laboratories, but also the outsourced ones.*		

wD56	**INDUSTRIAL DISCHARGES TESTS CARRIED OUT**	Unit: ***No.*** Valid values: Integer	**Input data** Used for indicators: *wOp53* Used for variables: none
	Number of industrial discharges tests carried out during the assessment period. *Includes not only tests performed in the undertaking laboratories, but also the outsourced ones.*		

wD57	**WASTEWATER QUALITY TESTS REQUIRED**	Unit: ***No.*** Valid values: Integer	**Input data or** **wD57 = wD58+wD59+wD60+wD61+ wD62+wD63+wD64** Used for indicators: *wOp44* Used for variables: none
	Total number of wastewater quality tests required by applicable standards or legislation, during the assessment period.		

wD58	**BOD TESTS REQUIRED**	Unit: ***No.*** Valid values: Integer	**Input data** Used for indicators: *wOp44*(wD57), *wOp45* Used for variables: wD57
	Number of BOD tests required by applicable standards or legislation, during the assessment period.		

wD59	**COD TESTS REQUIRED**	Unit: ***No.*** Valid values: Integer	**Input data** Used for indicators: *wOp44*(wD57), *wOp46* Used for variables: wD57
	Number of COD tests required by applicable standards or legislation, during the assessment period.		

wD60	**TSS TESTS REQUIRED**	Unit: ***No.*** Valid values: Integer	**Input data** Used for indicators: *wOp44*(wD57), *wOp47* Used for variables: wD57
	Number of TSS tests required by applicable standards or legislation, during the assessment period.		

wD61	**TOTAL PHOSPHORUS TESTS REQUIRED**	Unit: ***No.*** Valid values: Integer	**Input data** Used for indicators: *wOp44*(wD57), *wOp48* Used for variables: wD57
	Number of phosphorus tests required by applicable standards or legislation, during the assessment period.		

wD62	**TOTAL NITROGEN TESTS REQUIRED**	Unit: ***No.*** Valid values: Integer	**Input data** Used for indicators: *wOp44*(wD57), *wOp49* Used for variables: wD57
	Number of nitrogen tests required by applicable standards or legislation, assessing nitrogen and nitrogen compounds, during the assessment period.		

wD63	**FAECAL *E.COLI* TESTS REQUIRED**	Unit: ***No.*** Valid values: Integer	**Input data** Used for indicators: *wOp44*(wD57), *wOp50* Used for variables: wD57
	Number of faecal *Escherichia coli* tests required by applicable standards or legislation, during the assessment period.		

wD64	**OTHER WASTEWATER QUALITY TESTS REQUIRED**	Unit: ***No.*** Valid values: Integer	**Input data** Used for indicators: *wOp44*(wD57), *wOp51* Used for variables: wD57
	Number of other wastewater quality tests required by applicable standards or legislation, during the assessment period.		

wD65	**SLUDGE TESTS REQUIRED**	Unit: ***No.*** Valid values: Integer	**Input data**
	Number of *sludge* tests required by applicable standards or legislation, during the assessment period.		Used for indicators: *wOp52* Used for variables: none

wD66	**INDUSTRIAL DISCHARGES TESTS REQUIRED**	Unit: ***No.*** Valid values: Integer	**Input data**
	Number of industrial discharges tests required by applicable standards or legislation, during the assessment period.		Used for indicators: *wOp53* Used for variables: none

wD67	**PERMANENT VEHICLES**	Unit: ***No.*** Valid values: Integer	**Input data**
	Average daily number of vehicles available for *sewer system* related fieldwork, at the reference date. *In the case of multi-utilities, only the portion of vehicles related to wastewater service activities shall be considered. All vehicles that are available shall be accounted for, regardless of their ownership, undertaking or subcontractors. Vehicles used in the scope of outsourced activities shall be excluded. Heavy equipment shall not be included.*		Used for indicators: *wOp54* Used for variables: none

wD68	**GAS DETECTORS**	Unit: ***No.*** Valid values: Integer	**Input data**
	Number of gas detectors (including: O_2, H_2S, CH_4 and CO_2) at the reference date. *All gas detectors that are available shall be accounted for, regardless of their ownership, undertaking or subcontractors. Include both permanently installed and portable gas detectors that are calibrated.*		Used for indicators: *wOp55, wOp56* Used for variables: none

wD69	**PERMANENTLY INSTALLED GAS DETECTORS**	Unit: ***No.*** Valid values: Integer	**Input data**
	Number of permanently installed gas detectors (including: O_2, H_2S, CH_4 and CO_2) at the reference date.		Used for indicators: *wOp56* Used for variables: none

SECTION E - DEMOGRAPHY (AND CUSTOMER) DATA

wE1	**RESIDENT POPULATION**	Unit: ***Inhab.*** Valid values: Integer	**Input data**
	Total population living permanently in the area that is the responsibility of the wastewater undertaking, at the reference date.		Used for indicators: *wQS1, wQS2, wQS3, wQS4, wQS19, wQS20, wQS21, wQS22, wQS23, wQS24, wQS25, wQS26* Used for variables: none

wE2	**RESIDENT POPULATION SERVED BY WWTP**	Unit: ***Inhab.*** Valid values: Integer	**Input data**
	Resident population served by *wastewater treatment plants* managed by the undertaking, at the reference date.		Used for indicators: *wQS2* Used for variables: none

wE3	**RESIDENT POPULATION SERVED BY ON-SITE SYSTEMS**	Unit: ***Inhab.*** Valid values: Integer	**Input data** Used for indicators: *wQS3* Used for variables: none
	Resident population served by *on-site systems* managed by the undertaking (e.g. *septic tanks*, latrines), at the reference date.		

wE4	**RESIDENT POPULATION CONNECTED TO SE**	Unit: ***Inhab.*** Valid values: Integer	**Input data** Used for indicators: *wQS1* Used for variables: none
	Resident population connected to the *sewer systems* managed by the undertaking, at the reference date.		

wE5	**POPULATION EQUIVALENT SERVED BY WWTP**	Unit: ***p.e.*** Valid values: Integer	**Input data** Used for indicators: *wEn1, wEn6, wPe1, wPe10, wOp18* Used for variables: none
	population equivalent served by *wastewater treatment plants* managed by the undertaking, at the reference date. *This variable is estimated using the average BOD_5 loads of wastewater treatment plants inflow and the population equivalent definition (See Appendix 1 - Glossary).*		

wE6	**POPULATION EQUIVALENT SERVED BY ON-SITE SYSTEMS**	Unit: ***p.e.*** Valid values: Integer	**Input data** Used for indicators: *wEn15, wPe1, wPe10* Used for variables: none
	population equivalent served by *on-site systems* that are the responsibility of the undertaking, at the reference date. *This variable is estimated using the average BOD_5 loads of on-site systems inflow and the population equivalent definition (See Appendix 1 - Glossary).*		

wE7	**RESIDENT POPULATION NOT SERVED**	Unit: ***Inhab.*** Valid values: Integer	**Input data** Used for indicators: *wQS4* Used for variables: none
	Resident population whose wastewater is neither collected nor treated, at the reference date.		

wE8	**TOTAL POPULATION EQUIVALENT SERVED**	Unit: ***p.e.*** Valid values: Integer	**Input data** Used for indicators: *wFi1, wFi5, wFi7, wFi9, wFi27* Used for variables: none
	Total population producing wastewater that is either collected or treated by the undertaking at the reference date, including *imported wastewater* and industrial contributions expressed in *population equivalent.* *This variable is estimated using average BOD_5 loads, wastewater flows and the population equivalent definition (See Appendix 1 - Glossary).*		

SECTION F - QUALITY OF SERVICE DATA

wF1	**COLLECTED SEWAGE**	Unit: ***m³*** Valid values: ≥ 0	**Input data** Used for indicators: *wOp30, wQS5, wQS6, wQS7, wQS8, wQS9* Used for variables: none
	collected sewage, corresponding to the volume of domestic, commercial and industrial inputs to the *sewer system* during the assessment period. *This data should be obtained from a measure of dry weather flow subtracting infiltration, in dry weather conditions. In case there are no measurements of dry weather flow, collected sewage can be estimated based on a return factor, which is a percentage of water supplied that is discharged as wastewater.*		

wF2	**DRY WEATHER FLOODING OF PROPERTIES FROM SANITARY SEWERS**	Unit: ***No.*** Valid values: Integer	**Input data** Used for indicators: *wQS10* Used for variables: none
	Number of properties affected by *flooding* from *sanitary sewers* in *dry weather*, during the assessment period.		

wF3	**WET WEATHER FLOODING OF PROPERTIES FROM SANITARY SEWERS**	Unit: ***No.*** Valid values: Integer	**Input data** Used for indicators: *wQS11* Used for variables: none
	Number of properties affected by *flooding* from *sanitary sewers* in *wet weather*, during the assessment period.		

wF4	**DRY WEATHER FLOODING OF PROPERTIES FROM COMBINED SEWERS**	Unit: ***No.*** Valid values: Integer	**Input data** Used for indicators: *wQS12* Used for variables: none
	Number of properties affected by *flooding* from combined sewers in *dry weather*, during the assessment period.		

wF5	**WET WEATHER FLOODING OF PROPERTIES FROM COMBINED SEWERS**	Unit: ***No.*** Valid values: Integer	**Input data** Used for indicators: *wQS13* Used for variables: none
	Number of properties affected by *flooding* from combined sewers in *wet weather*, during the assessment period.		

wF6	**WET WEATHER SURFACE FLOODING OF PROPERTIES**	Unit: ***No.*** Valid values: Integer	**Input data** Used for indicators: *wQS14* Used for variables: none
	Number of properties affected by *surface flooding* in *wet weather*, during the assessment period.		

wF7	**WASTEWATER SERVICE INTERRUPTIONS**	Unit: ***Hours*** Valid values: Integer	**Input data** Used for indicators: *wQS15* Used for variables: none
	Sum, for the assessment period, of the number of properties affected by *service interruption* multiplied by the respective duration of interruptions in hours.		

wF8	**NEW CONNECTIONS ESTABLISHED**	Unit: ***No.*** Valid values: Integer	**Input data** Used for indicators: *wQS16* Used for variables: none
	Number of new *service connections* established during the assessment period. *This variable refers to new contracts, when service connections already exist.*		

wF9	**NEW SERVICE CONNECTIONS ESTABLISHMENT TIME**	Unit: ***Day*** Valid values: Real	**Input data** Used for indicators: *wQS16* Used for variables: none
	Total time spent for establishing new *service connections*, from the customer request until the availability of the service, during the assessment period.		

wF10	**CONNECTIONS REPAIRED**	Unit: ***No.*** Valid values: Integer	**Input data** Used for indicators: *wQS17* Used for variables: none
	Total number of connections *repaired* during the assessment period. *This variable includes all reported service connections failures beyond failures reported by customers.*		

wF11	**CONNECTIONS REPAIR TIME**	Unit: ***Day*** Valid values: Real	**Input data** Used for indicators: *wQS17* Used for variables: none
	Total time spent for *repairing service connections*, from the time failures are reported until the service is resumed, during the assessment period. *This variable includes all reported service connection failures beyond failures reported by customers.*		

wF12	**TOTAL COMPLAINTS**	Unit: ***No.*** Valid values: Integer	**Input data or** **wF12 = wF13+wF14+wF15+wF16+ wF17+wF18+wF19** Used for indicators: *wQS19*, *wQS27* Used for variables: none
	Total number of *complaints* relating to wastewater system performance, during the assessment period. *This variable includes all direct, telephone, and written complaints.*		

wF13	**BLOCKAGE COMPLAINTS**	Unit: ***No.*** Valid values: Integer	**Input data** Used for indicators: *wQS19*(wF12), *wQS20*, *wQS27*(wF12) Used for variables: wF12
	Number of *complaints* as a result of *blockages*, during the assessment period. *This variable includes all direct, telephone, and written complaints related to blockages.*		

wF14	**FLOODING COMPLAINTS**	Unit: ***No.*** Valid values: Integer	**Input data** Used for indicators: *wQS19*(wF12), *wQS21*, *wQS27*(wF12) Used for variables: wF12
	Number of *complaints* as a result of *flooding*, during the assessment period. *This variable includes all direct, telephone, and written complaints related to flooding occurrences.*		

wF15	**POLLUTION COMPLAINTS**	Unit: ***No.*** Valid values: Integer	**Input data** Used for indicators: *wQS19*(wF12), *wQS22*, *wQS27*(wF12) Used for variables: wF12
	Number of *complaints* as a result of pollution incidents, during the assessment period. *This variable includes all direct, telephone, and written complaints related to pollution incidents.*		

wF16	**ODOUR COMPLAINTS**	Unit: ***No.*** Valid values: Integer	**Input data** Used for indicators: *wQS19*(wF12), *wQS23*, *wQS27*(wF12) Used for variables: wF12
	Number of *complaints* as a result of odours, during the assessment period. *This variable includes all direct, telephone, and written complaints related to odours.*		

wF17	**RODENTS RELATED COMPLAINTS**	Unit: ***No.*** Valid values: Integer	**Input data** Used for indicators: *wQS19*(wF12), *wQS24*, *wQS27*(wF12) Used for variables: wF12
	Number of *complaints* as a result of rodents, during the assessment period. *This variable includes all direct, telephone, and written complaints related to rodents.*		

wF18	**CUSTOMER ACCOUNT RELATED COMPLAINTS**	Unit: ***No.*** Valid values: Integer	**Input data** Used for indicators: *wQS19*(wF12), *wQS25*, *wQS27*(wF12) Used for variables: wF12
	Number of *complaints* related to costumer accounts, during the assessment period. *This variable includes all direct, telephone, and written complaints related to customer accounts.*		

wF19	**OTHER COMPLAINTS**	Unit: ***No.*** Valid values: Integer	**Input data** Used for indicators: *wQS19*(wF12), *wQS26*, *wQS27*(wF12) Used for variables: wF12
	Number of other *complaints* during the assessment period. *This variable includes all direct, telephone, and written complaints due to other reasons.*		

wF20	**RESPONSES TO COMPLAINTS**	Unit: ***No.*** Valid values: Integer	**Input data** Used for indicators: *wQS27* Used for variables: none
	Number of responses to *complaints* given within the target time, during the assessment period. *Target time is the time specified by the wastewater service provider, by a regulator or other authorised agency, taken for the service provider to respond to a telephone, written, or other form of complaint, loss of service, or system fault.*		

wF21	**THIRD PARTY DAMAGES**	Unit: ***No.*** Valid values: Integer	**Input data** Used for indicators: *wQS28* Used for variables: none
	Number of occurrences, during the assessment period, related to accidents with persons not working for the undertaking or causing damage of third party properties, assets, vehicles or belongings.		

wF22	**ACCIDENTS**	Unit: ***No.*** Valid values: Integer	**Input data** Used for indicators: *wQS28* Used for variables: none
	Total number of accidents that are the undertaking responsibility, during the assessment period.		

wF23	**SEPTIC TANK EMPTYING REQUESTS**	Unit: ***No.*** Valid values: Integer	**Input data** Used for indicators: *wQS18* Used for variables: none
	Total number of requests for emptying *septic tanks* or contents of pit latrines, during the assessment period.		

wF24	**SEPTIC TANK EMPTYING RESPONSE TIME**	Unit: ***Day*** Valid values: Real	**Input data**
	Sum, for all requests, of the response time to empty *septic tanks* or contents of pit latrines, during the assessment period. *Response time is the period in days from request to empty septic tanks or contents of pit latrines to the day the service is rendered.*		Used for indicators: *wQS18* Used for variables: none

wF25	**TRAFFIC DISTURBANCES**	Unit: ***No.*** Valid values: Integer	**Input data**
	Number of traffic disturbances during the assessment period. *Traffic disturbances include activities related to inspection, cleaning, maintenance, construction and rehabilitation works in the system, where traffic is interrupted in one traffic lane at least. Roads where average traffic is below 5000 vehicles per day during the works shall not be considered.*		Used for indicators: *wQS29* Used for variables: none

wF26	**TRAFFIC LANE INTERRUPTIONS**	Unit: ***km x hour*** Valid values: Real	**Input data**
	$\Sigma_{\text{for all traffic disturbances due to wastewater related activities}}$ (Traffic lane length unavailable x interruption duration), during the assessment period. *Traffic disturbances include activities related to inspection, cleaning, maintenance, construction and rehabilitation works in the sewer system where traffic is interrupted in one traffic lane at least. Roads where average traffic is below 5000 vehicles per day during the works shall not be considered.*		Used for indicators: *wQS29* Used for variables: none

SECTION G – ECONOMIC AND FINANCIAL DATA

wG1	**TOTAL REVENUES**	Unit: ***US$*** Valid values: ≥ 0	**Input data or** **wG1 = wG2 - wG33**
	Total operating revenues (wG2) minus capitalised costs of self constructed assets (wG33), regarding the wastewater service, during the assessment period. *Capitalised costs of self-constructed assets have more correctly to be intended not as revenue but as an economic correction of each type of cost to negative apportion. These capitalisations have consequently to be negative allocated even for the calculation of annual revenues. See diagram of 4.6 - Financial aspects, Table 7 and Table 8. Exchange rate of local currencies shall be referred to the end of the year.*		Used for indicators: *wFi2, wFi3, wFi30, wFi31* Used for variables: none

wG2	**TOTAL OPERATING REVENUES**	Unit: ***US$*** Valid values: ≥ 0	**Input data**
	Total operating revenues, including service revenues (wG3), work in progress, capitalised costs of self-constructed assets (wG33) and other operating revenues, regarding the wastewater service, during the assessment period. *See diagram of 4.6 - Financial aspects, Table 7 and Table 8. Exchange rate of local currencies shall be referred to the end of the year.*		Used for indicators: *wFi1*, *wFi2*(wG1), *wFi3*(wG1), *wFi30*(wG1), *wFi31*(wG1), *wFi38* Used for variables: wG1

wG3	**SERVICE REVENUES** Operating revenues from wastewater services, during the assessment period. *See diagram of 4.6 - Financial aspects, Table 7 and Table 8. Exchange rate of local currencies shall be referred to the end of the year.*	Unit: ***US$*** Valid values: ≥ 0	**Input data** Used for indicators: *wFi2, wFi3, wFi4, wFi32, wFi45* Used for variables: none
wG4	**INDUSTRIAL REVENUES** Operating revenues from industries, during the assessment period. *See diagram of 4.6 - Financial aspects, Table 7 and Table 8. Exchange rate of local currencies shall be referred to the end of the year.*	Unit: ***US$*** Valid values: ≥ 0	**Input data** Used for indicators: *wFi4* Used for variables: none
wG5	**TOTAL COSTS** Total costs, including capital and running costs, regarding the wastewater service, during the assessment period. *See diagram of section 4.6 - Financial aspects, Table 7 and Table 8. Exchange rate of local currencies shall be referred to the end of the year.*	Unit: ***US$*** Valid values: ≥ 0	**Input data or** **wG5 = wG6+wG7** Used for indicators: *wFi5, wFi6, wFi30* Used for variables: none
wG6	**RUNNING COSTS** Total operation and maintenance costs and internal manpower costs, excluding the capitalised costs of self-constructed assets, regarding the wastewater service, during the assessment period. *This definition has, on aggregate level, to be equivalent to the sum of the NET desegregated values allocated at the numerator of the indicators figuring the composition of running costs per type of cost: wFi11, wFi12, wFi13 and wFi15. See diagram of section 4.6 - Financial aspects, Table 7 and Table 8. Exchange rate of local currencies shall be referred to the end of the year.*	Unit: ***US$*** Valid values: ≥ 0	**Input data or** **wG6 = wG8+wG9** Used for indicators: *wFi5, wFi6*(wG5), *wFi7, wFi8, wFi11, wFi12, wFi13, wFi14, wFi15, wFi16, wFi17, wFi18, wFi19, wFi20, wFi21, wFi22, wFi23, wFi24, wFi30*(wG5), *wFi31* Used for variables: wG5
wG7	**CAPITAL COSTS** Total annual net interest and depreciation (based on book values), regarding the wastewater service, during the assessment period. *The net value of interest has to be considered, being the interest income a reduction in capital costs and not a revenue. See section 4.6 - Financial aspects, Table 7 and Table 8. Exchange rate of local currencies shall be referred to the end of the year.*	Unit: ***US$*** Valid values: ≥ 0	**Input data or** **wG7 = wG26+wG29** Used for indicators: *wFi5, wFi6*(wG5), *wFi9, wFi10, wFi25, wFi26, wFi30*(wG5) Used for variables: wG5
wG8	**OPERATIONAL COSTS** Operational costs, regarding the wastewater service, including the net value (obtained by negative apportionment of related capitalised costs of self-constructed assets) of energy, external services, leasing and rentals, purchases, taxes, levies and fees, exceptional earnings and losses, other operating costs, excluding manpower, during the assessment period. *See section 4.6 - Financial aspects, Table 7 and Table 8. Exchange rate of local currencies shall be referred to the end of the year.*	Unit: ***US$*** Valid values: ≥ 0	**Input data or** **wG8 = wG10+wG11+ wG12+wG13+ wG14+wG15+wG16** Used for indicators: *wFi5*(wG6), *wFi6*(wG5), *wFi7*(wG6), *wFi8*(wG6), *wFi11*(wG6), *wFi12*(wG6), *wFi13*(wG6), *wFi14*(wG6), *wFi15*(wG6), *wFi16*(wG6), *wFi17*(wG6), *wFi18*(wG6), *wFi19*(wG6), *wFi20*(wG6), *wFi21*(wG6), *wFi22*(wG6), *wFi23*(wG6), *wFi24*(wG6), *wFi30*(wG5), *wFi31*(wG6) Used for variables: wG5(wG6), wG6

wG9	**INTERNAL MANPOWER COSTS**	Unit: ***US$*** Valid values: ≥ 0	**Input data**
	Internal manpower costs regarding the wastewater service, referred to the net value (obtained by negative apportionment of related capitalised costs of self-constructed assets) of the personnel which salary is paid directly by the undertaking, during the assessment period. *See section 4.6 - Financial aspects, Table 7 and Table 8. Manpower costs include salaries and other costs that derive directly from them, such as complementary payments or social security complements. Exchange rate of local currencies shall be referred to the end of the year.*		Used for indicators: *wFi5*(wG6), *wFi6*(wG5), *wFi7*(wG6), *wFi8*(wG6), *wFi11*, *wFi12*(wG6), *wFi13*(wG6), *wFi14*(wG6), *wFi15*(wG6), *wFi16*(wG6), *wFi17*(wG6), *wFi18*(wG6), *wFi19*(wG6), *wFi20*(wG6), *wFi21*(wG6), *wFi22*(wG6), *wFi23*(wG6), *wFi24*(wG6), *wFi30*(wG5), *wFi31*(wG6) Used for variables: wG5(wG6), wG6

wG10	**EXTERNAL SERVICES COSTS**	Unit: ***US$*** Valid values: ≥ 0	**Input data**
	The net value (obtained by negative apportionment of related capitalised costs of self-constructed assets) of total costs of external services (i.e. outsourcing), external manpower costs included, regarding the wastewater service, during the assessment period. *See section 4.6 - Financial aspects, Table 7 and Table 8. Exchange rate of local currencies shall be referred to the end of the year.*		Used for indicators: *wFi5*(wG6), *wFi6*(wG5), *wFi7*(wG6), *wFi8*(wG6), *wFi11*(wG6), *wFi12*, *wFi13*(wG6), *wFi14*(wG6), *wFi15*(wG6), *wFi16*(wG6), *wFi17*(wG6), *wFi18*(wG6), *wFi19*(wG6), *wFi20*(wG6), *wFi21*(wG6), *wFi22*(wG6), *wFi23*(wG6), *wFi24*(wG6), *wFi30*(wG5), *wFi31*(wG6) Used for variables: wG5(wG6), wG6(wG8), wG8

wG11	**ENERGY COSTS**	Unit: ***US$*** Valid values: ≥ 0	**Input data**
	Total cost of energy regarding the wastewater service, during the assessment period. *See diagram of section 4.6 - Financial aspects, Table 7 and Table 8. This variable includes not only the component proportional to the energy consumption but all the other components of energy costs, such as power tariffs and taxes. Energy consumption for other activities different of the wastewater pumping are also included. Exchange rate of local currencies shall be referred to the end of the year.*		Used for indicators: *wFi5*(wG6), *wFi6*(wG5), *wFi7*(wG6), *wFi8*(wG6), *wFi11*(wG6), *wFi12*(wG6), *wFi13*, *wFi14*(wG6), *wFi15*(wG6), *wFi16*(wG6), *wFi17*(wG6), *wFi18*(wG6), *wFi19*(wG6), *wFi20*(wG6), *wFi21*(wG6), *wFi22*(wG6), *wFi23*(wG6), *wFi24*(wG6), *wFi30*(wG5), *wFi31*(wG6) Used for variables: wG5(wG6), wG6(wG8), wG8

wG12	**PURCHASED MERCHANDISES**	Unit: ***US$*** Valid values: ≥ 0	**Input data**
	The net value (obtained by negative apportionment of related capitalised costs of self-constructed assets) of all materials, chemicals and other consumables not included in external services costs, regarding the wastewater service, during the assessment period. *See diagram of section 4.6 - Financial aspects, Table 7 and Table 8. Exchange rate of local currencies shall be referred to the end of the year.*		Used for indicators: *wFi5*(wG6), *wFi6*(wG5), *wFi7*(wG6), *wFi8*(wG6), *wFi11*(wG6), *wFi12*(wG6), *wFi13*(wG6), *wFi14*, *wFi15*(wG6), *wFi16*(wG6), *wFi17*(wG6), *wFi18*(wG6), *wFi19*(wG6), *wFi20*(wG6), *wFi21*(wG6), *wFi22*(wG6), *wFi23*(wG6), *wFi24*(wG6), *wFi30*(wG5), *wFi31*(wG6) Used for variables: wG5(wG6), wG6(wG8), wG8

wG13	**LEASING AND RENTALS**	Unit: ***US$*** Valid values: ≥ 0	**Input data**
	The net value (obtained by negative apportionment of related capitalised costs of self-constructed assets) of total cost of leasing and rentals, regarding the wastewater service, during the assessment period. *See diagram of section 4.6 - Financial aspects, Table 7 and Table 8. Exchange rate of local currencies shall be referred to the end of the year.*		Used for indicators: *wFi5*(wG6), *wFi6*(wG5), *wFi7*(wG6), *wFi8*(wG6), *wFi11*(wG6), *wFi12*(wG6), *wFi13*(wG6), *wFi14*(wG6), *wFi15*, *wFi16*(wG6), *wFi17*(wG6), *wFi18*(wG6), *wFi19*(wG6), *wFi20*(wG6), *wFi21*(wG6), *wFi22*(wG6), *wFi23*(wG6), *wFi24*(wG6), *wFi30*(wG5), *wFi31*(wG6) Used for variables: wG5(wG6), wG6(wG8), wG8

wG14	**TAXES, LEVIES AND FEES**	Unit: ***US$*** Valid values: ≥ 0	**Input data**
	All levies and licences strictly connected with plant's operation to be paid to a governmental or municipal authority (excluding direct taxes on EBT), regarding the wastewater service, during the assessment period. *See diagram of section 4.6 - Financial aspects, Table 7 and Table 8. Exchange rate of local currencies shall be referred to the end of the year.*		Used for indicators: *wFi5*(wG6), *wFi6*(wG5), *wFi7*(wG6), *wFi8*(wG6), *wFi11*(wG6), *wFi12*(wG6), *wFi13*(wG6), *wFi14*(wG6), *wFi15*, *wFi16*(wG6), *wFi17*(wG6), *wFi18*(wG6), *wFi19*(wG6), *wFi20*(wG6), *wFi21*(wG6), *wFi22*(wG6), *wFi23*(wG6), *wFi24*(wG6), *wFi30*(wG5), *wFi31*(wG6) Used for variables: wG5(wG6), wG6(wG8), wG8

wG15	**EXCEPTIONAL EARNINGS AND LOSSES**	Unit: ***US$*** Valid values: Real	**Input data**
	Exceptional income or expenditures related to sales / writing-off of fixed assets, regarding the wastewater service, during the assessment period. *See diagram of section 4.6 - Financial aspects, Table 7 and Table 8. Exchange rate of local currencies shall be referred to the end of the year.*		Used for indicators: *wFi5*(wG6), *wFi6*(wG5), *wFi7*(wG6), *wFi8*(wG6), *wFi11*(wG6), *wFi12*(wG6), *wFi13*(wG6), *wFi14*(wG6), *wFi15*, *wFi16*(wG6), *wFi17*(wG6), *wFi18*(wG6), *wFi19*(wG6), *wFi20*(wG6), *wFi21*(wG6), *wFi22*(wG6), *wFi23*(wG6), *wFi24*(wG6), *wFi30*(wG5), *wFi31*(wG6) Used for variables: wG5(wG6), wG6(wG8), wG8

wG16	**OTHER OPERATING COSTS**	Unit: ***US$*** Valid values: ≥ 0	**Input data**
	The net value (obtained by negative apportionment of related capitalised costs of self-constructed assets) of other operating costs, not included in chemicals, other consumables and materials for maintenance and *repair*, leasing and rentals, taxes, levies & fees and exceptional earnings & losses, regarding the wastewater service, during the assessment period. *See diagram of section 4.6 - Financial aspects, Table 7 and Table 8. Exchange rate of local currencies shall be referred to the end of the year.*		Used for indicators: *wFi5*(wG6), *wFi6*(wG5), *wFi7*(wG6), *wFi8*(wG6), *wFi11*(wG6), *wFi12*(wG6), *wFi13*(wG6), *wFi14*(wG6), *wFi15*, *wFi16*(wG6), *wFi17*(wG6), *wFi18*(wG6), *wFi19*(wG6), *wFi20*(wG6), *wFi21*(wG6), *wFi22*(wG6), *wFi23*(wG6), *wFi24*(wG6), *wFi30*(wG5), *wFi31*(wG6) Used for variables: wG5(wG6), wG6(wG8), wG8

wG17	**GENERAL MANAGEMENT RUNNING COSTS**	Unit: ***US$*** Valid values: ≥ 0	**Input data**
	Part of the running costs related to the net value (obtained by negative apportionment of related capitalised costs of self-constructed assets) of costs of directorate, central administration, strategic planning, public relations, legal affairs, internal audits, environmental policy and general computing support, regarding the wastewater service, during the assessment period. *This variable includes not only internal functions costs but also, when applicable, outsourced functions costs. See section 4.5 - Organisational functions and definitions, Figure 5 and Table 1. Exchange rate of local currencies shall be referred to the end of the year.*		Used for indicators: *wFi16* Used for variables: none

wG18	**HUMAN RESOURCES MANAGEMENT RUNNING COSTS**	Unit: ***US$*** Valid values: ≥ 0	**Input data** Used for indicators: *wFi17* Used for variables: none
	Part of the running costs related to the net value (obtained by negative apportionment of related capitalised costs of self-constructed assets) of costs of personnel administration, education and training, occupational safety and medicine services and social activities, regarding the wastewater service, during the assessment period. *This variable includes not only internal functions costs but also, when applicable, outsourced functions costs. See section 4.5 - Organisational functions and definitions, Figure 5 and Table 2. Exchange rate of local currencies shall be referred to the end of the year.*		

wG19	**FINANCIAL AND COMMERCIAL MANAGEMENT RUNNING COSTS**	Unit: ***US$*** Valid values: ≥ 0	**Input data** Used for indicators: *wFi18* Used for variables: none
	Part of the running costs related to the net value (obtained by negative apportionment of related capitalised costs of self-constructed assets) of costs of economic and financial planning, economic administration, economic controlling and purchasing and material management, regarding the wastewater service, during the assessment period. *This variable includes not only internal functions costs but also, when applicable, outsourced functions costs. See section 4.5 - Organisational functions and definitions, Figure 5 and Table 3. Exchange rate of local currencies shall be referred to the end of the year.*		

wG20	**CUSTOMER SERVICE MANAGEMENT RUNNING COSTS**	Unit: ***US$*** Valid values: ≥ 0	**Input data** Used for indicators: *wFi19* Used for variables: none
	Part of the running costs related to the net value (obtained by negative apportionment of related capitalised costs of self-constructed assets) of costs of accounting and control, customer relations and management, regarding the wastewater service, during the assessment period. *This variable includes not only internal functions costs but also, when applicable, outsourced functions costs. See definitions of section 4.5 - Organisational functions and definitions, Figure 5 and Table 4. Exchange rate of local currencies shall be referred to the end of the year.*		

wG21	**TECHNICAL MANAGEMENT COSTS**	Unit: ***US$*** Valid values: ≥ 0	**Input data** Used for indicators: *wFi20* Used for variables: none
	Part of the running costs related to the net value (obtained by negative apportionment of related capitalised costs of self-constructed assets) of wastewater service technical management (planning, construction, operation and maintenance), during the assessment period. *This variable includes not only internal functions costs but also, when applicable, outsourced functions costs. See definitions of section 4.5 - Organisational functions and definitions, Table 5 and Table 6. Exchange rate of local currencies shall be referred to the end of the year.*		

wG22	**WASTEWATER TREATMENT RUNNING COSTS**	Unit: ***US$*** Valid values: ≥ 0	**Input data** Used for indicators: *wFi21* Used for variables: none
	Part of the running costs related to the net value (obtained by negative apportionment of related capitalised costs of self-constructed assets) of wastewater treatment, during the assessment period. *For the ease of assessment, this variable can be broken down into more detailed sub-variables.*		

wG23	**SEWER SYSTEM RUNNING COSTS**	Unit: ***US$*** Valid values: ≥ 0	**Input data** Used for indicators: *wFi22* Used for variables: none
	Part of running costs related to the net value (obtained by negative apportionment of related capitalised costs of self-constructed assets) of *sewer system*, during the assessment period. *For the ease of assessment, this variable can be broken down into more detailed sub-variables.*		

wG24	**WASTEWATER QUALITY MONITORING RUNNING COSTS**	Unit: ***US$*** Valid values: ≥ 0	**Input data** Used for indicators: *wFi23* Used for variables: none
	Part of the running costs related to the net value (obtained by negative apportionment of related capitalised costs of self-constructed assets) of wastewater quality monitoring, during the assessment period. *In the case of multi-utilities, each employee shall be accounted for based on the proportion of time dedicated to wastewater related activities.* *For the ease of assessment, this variable can be broken down into more detailed sub-variables.*		

wG25	**SUPPORT SERVICES RUNNING COSTS**	Unit: ***US$*** Valid values: ≥ 0	**Input data** Used for indicators: *wFi24* Used for variables: none
	Part of the running costs related to the net value (obtained by negative apportionment of related capitalised costs of self-constructed assets) of support services, during the assessment period. *In the case of multi-utilities, each employee shall be accounted for based on the proportion of time dedicated to wastewater related activities.* *For the ease of assessment, this variable can be broken down into more detailed sub-variables.*		

wG26	**DEPRECIATION COSTS**	Unit: ***US$*** Valid values: ≥ 0	**Input data** Used for indicators: *wFi5*(wG7), *wFi6*(wG5), *wFi9*(wG7), *wFi10*(wG7), *wFi25*, *wFi26*(wG7), *wFi30*(wG5), *wFi33*, *wFi36* Used for variables: wG5(wG7), wG7
	Depreciation costs (on book values), regarding the wastewater service, during the assessment period. *See diagram of section 4.6 - Financial aspects, Table 7 and Table 8. Exchange rate of local currencies shall be referred to the end of the year.*		

wG27	**INTEREST EXPENSES COSTS**	Unit: ***US$*** Valid values: ≥ 0	**Input data** Used for indicators: *wFi5*(wG7), *wFi6*(wG5), *wFi9*(wG7), *wFi10*(wG7), *wFi25*(wG7), *wFi26*, *wFi30*(wG5) Used for variables: wG5(wG7), wG7(wG29), wG29
	Interest expenses costs, regarding the wastewater service, during the assessment period. *See section 4.6 - Financial aspects, Table 7 and Table 8. Exchange rate of local currencies shall be referred to the end of the year.*		

wG28	**INTEREST INCOME**	Unit: ***US$*** Valid values: ≥ 0	**Input data** Used for indicators: *wFi5*(wG7), *wFi6*(wG5), *wFi9*(wG7), *wFi10*(wG7), *wFi25*(wG7), *wFi26*, *wFi30*(wG5) Used for variables: wG5(wG7), wG7(wG29), wG29
	Total interest income, regarding the wastewater service, during the assessment period. *Interest income has more correctly to be intended not as revenue but as an economic correction of interest expenses. It has consequently to be negative allocated in order to obtain the net interest costs. See section 4.6 - Financial aspects, Table 7 and Table 8. Exchange rate of local currencies shall be referred to the end of the year.*		

wG29	**NET INTEREST**	Unit: ***US$*** Valid values: Real	**Input data or** **wG29 = wG27 - wG28** Used for indicators: *wFi5*(wG7), *wFi6*(wG5), *wFi9*(wG7), *wFi10*(wG7), *wFi25*(wG7), *wFi26*, *wFi30*(wG5) Used for variables: wG5(wG7), wG7
	Interest expenses costs minus interest income, regarding the wastewater service, during the assessment period. *Most times interest expenses are higher than income, what means that the net interest is a cost. See section 4.6 - Financial aspects, Table 7 and Table 8. Exchange rate of local currencies shall be referred to the end of the year.*		

wG30	**INVESTMENT IN TANGIBLE ASSETS**	Unit: ***US$*** Valid values: ≥ 0	**Input data or** **wG30 = wG31+wG32** Used for indicators: *wFi27*, *wFi28*, *wFi29*, *wFi34* Used for variables: none
	Total cost of the investments in tangible assets (expenditures for sewers, treatment plants and equipment), including capitalised cost of self-constructed tangible assets (apportionment of wG33 as for related to tangible assets), regarding the wastewater service, during the assessment period. *See section 4.6 - Financial aspects. Tangible assets include investment for supporting buildings, vehicles, etc. Exchange rate of local currencies shall be referred to the end of the year.*		

wG31	**INVESTMENTS FOR NEW ASSETS AND REINFORCEMENT OF EXISTING ASSETS**	Unit: ***US$*** Valid values: ≥ 0	**Input data** Used for indicators: *wFi27*(wG30), *wFi28*, *wFi29*(wG30), *wFi34*(wG30) Used for variables: wG30
	Total cost of the investments in tangible assets that constitute a new development for the service (new assets and *reinforcement* of existing assets), including capitalised cost of self-construction (apportionment of wG33 as for related to tangible new assets), regarding the wastewater service, during the assessment period. *Exchange rate of local currencies shall be referred to the end of the year.*		

wG32	**INVESTMENTS FOR ASSET REPLACEMENT AND RENOVATION**	Unit: ***US$*** Valid values: ≥ 0	**Input data** Used for indicators: *wFi27*(wG30), *wFi28*(wG30), *wFi29*, *wFi34*(wG30) Used for variables: wG30
	Total cost of the investments related to the *replacement* and *renovation* of existing assets ("like for like", i.e., maintaining approximately the same functionality of existing infrastructure), including capitalised cost of self-construction (apportionment of wG33 as for related to the *refurbishment* or replacement of tangible assets), regarding the wastewater service, during the assessment period. *Exchange rate of local currencies shall be referred to the end of the year.*		

wG33	**CAPITALISED COSTS OF SELF-CONSTRUCTED ASSETS**	Unit: ***US$*** Valid values: ≥ 0	**Input data** Used for indicators: *wFi1*, *wFi2*(wG1), *wFi3*(wG1), *wFi30*(wG1), *wFi31*(wG1) Used for variables: wG1
	Total capitalised cost of self-constructed assets, regarding the wastewater service, during the assessment period. *This variable includes capitalised cost of self-constructed assets regarding the investments both in tangible and in intangible assets that constitute a new development for the service and the investments for the replacement of existing assets. On these terms it represents: 1) an increase of the investments externally constructed and therefore included in wG2 and other connected data; 2) an economic correction of the gross value of Operation & Maintenance and Internal Manpower costs allocated in the Profit and Loss Statement 4.6 - Financial aspects, Table 7) and therefore negative apportioned to each type of these costs in wG8, wG9 and other connected data; 3) an economic correction of Total operating revenues allocated in the Profit and Loss Statement and therefore negative apportioned to obtain the Annual Revenues in wG1. See section 4.6 - Financial aspects, Table 7 and Table 8. Exchange rate of local currencies shall be referred to the end of the year.*		

wG34	**ACCOUNTS RECEIVABLE**	Unit: ***US$*** Valid values: ≥ 0	**Input data** Used for indicators: *wFi32* Used for variables: none
	Accounts receivable, regarding the wastewater service, at the reference date. *Exchange rate of local currencies shall be referred to the end of the year.*		

wG35	**INVESTMENTS SUBJECT TO DEPRECIATION**	Unit: ***US$*** Valid values: ≥ 0	**Input data** Used for indicators: *wFi33* Used for variables: none
	Investments for assets for which the general accepted accounting principles demands to practice depreciation, regarding the wastewater service, during the assessment period. *See diagram of section 4.6 - Financial aspects, Table 9. Exchange rate of local currencies shall be referred to the end of the year.*		

wG36	**INVESTMENTS FINANCED BY THE CASH-FLOW**	Unit: ***US$*** Valid values: ≥ 0	**Input data** Used for indicators: *wFi34* Used for variables: none
	Investments financed by the cash-flow (net income + depreciation + working capital variation), regarding the wastewater service, during the assessment period. *See diagram of section 4.6 - Financial aspects, Table 10. Exchange rate of local currencies shall be referred to the end of the year.*		

wG37	**DEPRECIATED HISTORICAL VALUE OF TANGIBLE ASSETS**	Unit: ***US$/year*** Valid values: ≥ 0	**Input data** Used for indicators: *wFi35*, *wFi42* Used for variables: none
	Sum, over all the tangible assets, of the depreciation value applied, during the year, since the date of purchase, regarding the wastewater service. *See diagram of section 4.6 - Financial aspects, Table 9 and Table 10. Exchange rate of local currencies shall be referred to the end of the year.*		

wG38	**HISTORICAL VALUE OF TANGIBLE ASSETS**	Unit: ***US$/year*** Valid values: ≥ 0	**Input data** Used for indicators: *wFi35, wFi36, wFi42* Used for variables: none
	Refers to the gross book value of total expenditures during the year (that appear for the net value in Table 11 in infrastructure and non-infrastructure tangible assets, regarding the wastewater service. *See diagram of section 4.6 - Financial aspects, Table 9 and Table 10. Exchange rate of local currencies shall be referred to the end of the year.*		

wG39	**ANNUAL DEBT FROM CUSTOMERS**	Unit: ***US$/year*** Valid values: ≥ 0	**Input data** Used for indicators: *wFi37* Used for variables: none
	Annual debt from customers, regarding the wastewater service. *Exchange rate of local currencies shall be referred to the end of the year.*		

wG40	**ANNUAL AMOUNT BILLED PER YEAR**	Unit: ***US$/year*** Valid values: ≥ 0	**Input data** Used for indicators: *wFi37* Used for variables: none
	Annual amount billed, regarding the wastewater service. *Exchange rate of local currencies shall be referred to the end of the year.*		

wG41	**CASH-FLOW**	Unit: ***US$*** Valid values: ≥ 0	**Input data** Used for indicators: *wFi39* Used for variables: none
	Total available is the sum of net income, depreciation and the net value of decrease or increase in working capital, regarding the wastewater service, during the assessment period. *See diagram of section 4.6 - Financial aspects, Table 10. Exchange rate of local currencies shall be referred to the end of the year.*		

wG42	**FINANCIAL DEBT SERVICE**	Unit: ***US$*** Valid values: ≥ 0	**Input data** Used for indicators: *wFi39* Used for variables: none
	The financial debt service contains the expenditures for interest (wG27), the cost of loans, and the principal (= capital) repayment debt instruments, regarding the wastewater service, during the assessment period. *Exchange rate of local currencies shall be referred to the end of the year.*		

wG43	**TOTAL DEBT**	Unit: ***US$/year*** Valid values: ≥ 0	**Input data or** **wG43 = wG48+wG49** Used for indicators: *wFi40, wFi43*(wG44) Used for variables: wG44
	Sum of long term liabilities (bonds and long term financial debts) and current liabilities, at the end of the fiscal year, regarding the wastewater service. *Exchange rate of local currencies shall be referred to the end of the year.*		

wG44	**SHAREHOLDERS' EQUITY**	Unit: ***US$/year*** Valid values: Real	**Input data or** **wG44 = wG46 - wG43** Used for indicators: *wFi40, wFi43* Used for variables: none
	Surplus of the asset over the liabilities, regarding the wastewater service, at the end of the fiscal year. *shareholders'* equity includes subscribed share capital, capital reserves, other reserves and net income for the year. *See diagram of section 4.6 - Financial aspects, Table 11. Exchange rate of local currencies shall be referred to the end of the year.*		

wG45	**CURRENT ASSETS** Current assets include cash at bank and in hand, accounts receivable from wastewater services, other accounts receivable, inventories and prepaid expenses, regarding the wastewater service, at the reference date. *See diagram of section 4.6 - Financial aspects, Table 11. Exchange rate of local currencies shall be referred to the end of the year.*	Unit: ***US$*** Valid values: ≥ 0	**Input data** Used for indicators: *wFi41* Used for variables: none

wG46	**TOTAL ASSETS** Sum of intangible assets (including goodwill and net value of licences and rights), tangible assets (including net value of wastewater undertaking plants and net value of other assets), financial assets (including net value of financial investments) and current assets (wG45), regarding the wastewater service, during the year. *See diagram of section 4.6 - Financial aspects, Table 11. Exchange rate of local currencies shall be referred to the end of the year.*	Unit: ***US$/year*** Valid values: ≥ 0	**Input data** Used for indicators: *wFi40*(wG44), *wFi43*(wG44), *wFi44*, *wFi45* Used for variables: wG44

wG47	**INVENTORIES** Accounted value of materials to be used in the production process that are in stock in the company, at the end of the fiscal year, regarding the wastewater service. *See diagram of section 4.6 - Financial aspects, Table 11. Exchange rate of local currencies shall be referred to the end of the year. This variable is a subset of wG45.*	Unit: ***US$*** Valid values: ≥ 0	**Input data** Used for indicators: *wFi38* Used for variables: none

wG48	**LONG TERM LIABILITIES** Sum of bonds and long term financial debts, at the end of the fiscal year, regarding the wastewater service. *Exchange rate of local currencies shall be referred to the end of the year.*	Unit: ***US$/year*** Valid values: ≥ 0	**Input data** Used for indicators: *wFi40*(wG43), *wFi43*(wG44) Used for variables: wG43, wG44(wG43)

wG49	**CURRENT LIABILITIES** Current liabilities include accounts payable, current portion of long term debt and miscellaneous current liabilities, regarding the wastewater service, at the reference date. *See diagram of section 4.6 - Financial aspects, Table 11. Exchange rate of local currencies shall be referred to the end of the year.*	Unit: ***US$*** Valid values: ≥ 0	**Input data** Used for indicators: *wFi40*(wG43), *wFi41*, *wFi43*(wG44) Used for variables: wG43, wG44(wG43)

wG50	**OPERATING INCOME** Earning before interest and taxes (EBIT), regarding the wastewater service, during the year. *See diagram of section 4.6 - Financial aspects, Table 11. Exchange rate of local currencies shall be referred to the end of the year.*	Unit: ***US$/year*** Valid values: ≥ 0	**Input data** Used for indicators: *wFi42*, *wFi44* Used for variables: none

wG51	**OPERATING INCOME RELATED TAXES**	Unit: ***US$/year*** Valid values: Real	**Input data**
	Subset of the total taxes specifically related to the operating income (EBIT), regarding the wastewater service, at the end of the fiscal year. *See diagram of section 4.6 - Financial aspects, Table 11. Exchange rate of local currencies shall be referred to the end of the year.*		Used for indicators: *wFi44* Used for variables: none

wG52	**NET INCOME**	Unit: ***US$/year*** Valid values: Real	**Input data**
	Net income after interest payment and taxes, regarding the wastewater service, at the end of the fiscal year. *See diagram of section 4.6 - Financial aspects. Exchange rate of local currencies shall be referred to the end of the year.*		Used for indicators: *wFi43* Used for variables: none

SECTION H - TIME DATA

wH1	**ASSESSMENT PERIOD**	Unit: ***days*** Valid values: Integer	**Input data**
	Period of time adopted for the assessment of data (variables, PI and CI). *The IWA PI system aims to be used annually and therefore it is highly recommended that the year is used as the reference assessment period. However, since the undertakings may need to track the evolution of their performance within the year, the PI system is prepared to accommodate other assessment periods for most indicators.* *In this case, and in order to ensure unit coherence and allow for PI comparison, all the PI expressed in terms of time are formulated in such a way that the values calculated for other assessment periods are converted into annual values.* *Attention is drawn to the fact that the behaviour of most variables is not uniform during the year, due to random or seasonal effects, or to activity planning. All comparisons based on PI assessed from non-annual data must take this fact into consideration, in order to avoid any bias.*		Used for indicators: *wEn3, wEn4, wEn6, wEn12, wEn13, wEn14, wEn15, wPe12, wPe17, wPe20, wPe21, wPe22, wPe23, wPe24, wPh8, wPh9, wOp1, wOp2, wOp3, wOp4, wOp5, wOp6, wOp7, wOp8, wOp9, wOp10, wOp11, wOp12, wOp13, wOp14, wOp15, wOp17, wOp18, wOp21, wOp22, wOp23, wOp24, wOp25, wOp26, wOp27, wOp28, wOp29, wOp31, wOp32, wOp33, wOp34, wOp35, wOp36, wOp37, wOp38, wOp39, wOp40, wOp41, wOp42, wOp44, wOp45, wOp46, wOp47, wOp48, wOp49, wOp50, wOp51, wOp52, wOp53, wQS10, wQS11, wQS12, wQS13, wQS14, wQS15, wQS17, wQS19, wQS20, wQS21, wQS22, wQS23, wQS24, wQS25, wQS26, wQS29, wFi1, wFi5, wFi6, wFi7, wFi8, wFi9, wFi10, wFi27, wFi32* Used for variables: none

REFERENCES

ASOTT (2002). **"Confined Spaces - What is a confined space?".** Retrieved February 17, 2003, from Australian School of Outdoor Technical Training Web site: http://www.asott.nsw.edu.au/confwk.htm

EN 752-1 (1995). **Drain and sewer systems outside the buildings – Part 1: Generalities and definitions.** European Committee for Standardisation (CEN).

Appendix 3 – Example of significance levels assignment for PI

RELATIVE SIGNIFICANCE LEVELS OF PI FOR A TYPICAL WASTEWATER UNDERTAKING IN A DEVELOPED COUNTRY

Table 16 provides a fictitious example of the relative significance of each of the PIs to a typical wastewater undertaking in a developed country. It is for a system with a high degree of sewer connectivity (>95%), treatment of the majority of sanitary wastewater (>95% to secondary quality), the presence of a number of CSOs and high quality sludge handling and disposal.

Table 16. Relative importance of each PI proposed from the wastewater undertaking point of view.

Group of indicators	code	Level 1 *L1*	Level 2 *L2*	Level 3 *L3*	Total
Environmental indicators	*wEn*	3	3	9	**15**
Personnel indicators	*wPe*	2	10	13	**25**
Physical indicators	*wPh*	1	6	5	**12**
Operational indicators	*wOp*	5	23	28	**56**
Quality of service indicators	*wQS*	6	13	10	**29**
Economic and financial indicators	*wFi*	8	15	22	**45**
Total number of indicators		**25**	**70**	**87**	**182**

ENVIRONMENTAL INDICATORS (wEn)

SUBGROUP	Code	Indicator	(unit)	Level
WASTEWATER	*wEn1*	WWTP compliance with discharge consents	(%)	L1
	wEn2	Wastewater reuse	(%)	L2
	wEn3	*intermittent overflow discharge* frequency	(No./overflow device/year)	L2
	wEn4	*intermittent overflow discharge* volume	(m3/overflow device/year)	L3
	wEn5	*intermittent overflow discharge* related to rainfall	(%/year)	L3
SOLID RESIDUES	*wEn6*	*sludge* production in WWTP	(kg DS/ p.e./year)	L1
	wEn7	*sludge utilisation*	(%)	L1
	wEn8	*sludge disposed*	(%)	L2
	wEn9	- *sludge* going to landfill	(%)	L3
	wEn10	- *sludge* thermally processed	(%)	L3
	wEn11	- other *sludge disposal*	(%)	L3
	wEn12	*sediments* from *sewers*	(ton/km sewer/year)	L3
	wEn13	*sediments* from *ancillaries*	(ton/km sewer/year)	L3
	wEn14	Solid waste from *screens* and grit	(ton/km sewer/year)	L3
	wEn15	*sediments* from *on-site systems*	(ton/p.e./year)	L3

PERSONNEL INDICATORS (wPe)

SUBGROUP	*Code*	Indicator	*(unit)*	*Level*
TOTAL PERSONNEL	*wPe1*	Personnel in WWT per *population equivalent*	*(No./ 10000 p.e.)*	*L1*
	wPe2	Personnel in SE per length of sewer	*(No./100 km sewer)*	*L1*
PERSONNEL PER MAIN FUNCTION	*wPe3*	General management personnel	*(%)*	*L2*
	wPe4	Human resources management personnel	*(%)*	*L2*
	wPe5	Financial and commercial services personnel	*(%)*	*L2*
	wPe6	Customer service personnel	*(%)*	*L2*
	wPe7	Technical services personnel	*(%)*	*L2*
	wPe8	- planning, design and construction personnel	*(%)*	*L3*
	wPe9	- operation and maintenance personnel	*(%)*	*L3*
TECHNICAL PERSONNEL PER ACTIVITY	*wPe10*	Technical WWT personnel	*(No./1000 p.e.)*	*L2*
	wPe11	Technical SE personnel	*(No./100 km sewer)*	*L2*
	wPe12	Wastewater quality monitoring personnel	*(No./(1000 tests/year))*	*L3*
	wPe13	Support services personnel	*(%)*	*L3*
PERSONNEL QUALIFICATION	*wPe14*	University degree personnel	*(%)*	*L3*
	wPe15	Basic education personnel	*(%)*	*L3*
	wPe16	Other personnel	*(%)*	*L3*
PERSONNEL TRAINING	*wPe17*	Total training of personnel	*(hours/employee/year)*	*L3*
PERSONNEL VACCINATION AND SAFETY	*wPe19*	Authorised confined spaces training	*(%)*	*L3*
	wPe18	Vaccination	*(%)*	*L3*
	wPe20	Working accidents	*(No./100 employees/year)*	*L2*
	wPe21	- personnel working fatalities	*(No./100 employees/year)*	*L2*
	wPe22	Absenteeism	*(days/100 employees/year)*	*L3*
	wPe23	- absenteeism due to accidents or illness at work	*(days/100 employees/year)*	*L3*
	wPe24	- absenteeism due to other reasons	*(days/100 employees/year)*	*L3*
OVERTIME WORK	*wPe25*	Overtime work	*(%)*	*L2*

PHYSICAL INDICATORS (wPh)

SUBGROUP	*Code*	Indicator	*(unit)*	*Level*
WASTEWATER TREATMENT	*wPh1*	*preliminary treatment* utilisation	*(%)*	*L2*
	wPh2	*primary treatment* utilisation	*(%)*	*L2*
	wPh3	*secondary treatment* utilisation	*(%)*	*L2*
	wPh4	*tertiary treatment* utilisation	*(%)*	*L2*
SEWERS	*wPh5*	Surcharging in gravity sewers in *dry weather*	*(%)*	*L1*
	wPh6	Surcharging in gravity sewers in *wet weather*	*(%)*	*L2*
	wPh7	High sewer *surcharging*	*(%)*	*L3*
PUMPING HEADROOM	*wPh8*	Pump power utilised in SE	*(%)*	*L2*
	wPh9	Pump power utilised in WWTP	*(%)*	*L3*
	wPh10	*sewer system* pump headroom	*(%)*	*L3*
AUTOMATION AND CONTROL	*wPh11*	Automation degree	*(%)*	*L3*
	wPh12	Remote control degree	*(%)*	*L3*

OPERATIONAL INDICATORS (wOp)

SUBGROUP	*Code*	Indicator	*(unit)*	*Level*
SEWER INSPECTION AND MAINTENANCE	*wOp1*	Sewer *inspection*	*(%/year)*	*L2*
	wOp2	Sewer *cleaning*	*(%/year)*	*L2*
	wOp3	Manhole chamber *inspection*	*(-/year)*	*L2*
	wOp4	*gully pots inspection*	*(-/year)*	*L2*
	wOp5	*gully pots cleaning*	*(-/year)*	*L2*
TANKS AND CSOs INSPECTION AND MAINTENANCE	*wOp6*	Tanks and *CSOs inspection* frequency	*(No./tank and CSO/year)*	*L2*
	wOp7	Tank and *CSOs inspection* volume	*(-/year)*	*L2*
	wOp8	Tank and *CSOs cleaning*	*(-/year)*	*L2*
	wOp9	*screen inspection*	*(-/year)*	*L3*
PUMPS AND PUMPING STATIONS INSPECTION	*wOp10*	*pumping station inspection* frequency	*(-/year)*	*L2*
	wOp11	Pump *inspection* by power	*(-/year)*	*L2*
EQUIPMENT CALIBRATION	*wOp12*	SE flow meters *calibration*	*(-/year)*	*L3*
	wOp13	WWTP flow meters *calibration*	*(-/year)*	*L3*
	wOp14	Wastewater quality monitoring equipment *calibration*	*(-/year)*	*L3*
ELECTRICAL AND SIGNAL TRANSMISSION EQUIPMENT INSPECTION	*wOp15*	Emergency power system *inspection*	*(-/year)*	*L3*
	wOp16	Signal transmission equipment *inspection*	*(-/year)*	*L3*
	wOp17	Electrical switchgear equipment *inspection*	*(-/year)*	*L3*
ENERGY CONSUMPTION	*wOp18*	WWT energy consumption	*(kWh/p.e./year)*	*L2*
	wOp19	WWT energy recovery from co-generation processes	*(%)*	*L3*
	wOp20	Standardised energy consumption	*(kWh/ m^3/1 m)*	*L3*
SEWER SYSTEM REHABILITATION	*wOp21*	Sewer *rehabilitation*	*(%/year)*	*L1*
	wOp22	- sewer *renovation*	*(%/year)*	*L3*
	wOp23	- sewer *replacement*	*(%/year)*	*L3*
	wOp24	- sewer and joints *repair*	*(No./100 km sewer/year)*	*L3*
	wOp25	Manhole chambers *replacement, renewal, renovation* or *repair*	*(%/year)*	*L2*
	wOp26	*replacement* of manhole covers	*(%/year)*	*L3*
	wOp27	*service connection rehabilitation*	*(%/year)*	*L2*
PUMPS REHABILITATION	*wOp28*	Pump *refurbishment*	*(%/year)*	*L2*
	wOp29	Pump *replacement*	*(%/year)*	*L2*
INFLOW / INFILTRATION / EXFILTRATION (I/I/E)	*wOp30*	*inflow / infiltration / exfiltration* (I/I/E)	*(%)*	*L2*
	wOp31	*inflow*	*(m^3/km/year)*	*L3*
	wOp32	*infiltration*	*(m^3/km/year)*	*L3*
	wOp33	*exfiltration*	*(m^3/km/year)*	*L3*
FAILURES	*wOp34*	Sewer *blockages*	*(No./100 km sewer/year)*	*L2*
	wOp35	- Sewer *blockage* locations	*(No./100 km sewer/year)*	*L3*
	wOp36	*pumping station blockages*	*(No./pumping station/year)*	*L2*
	wOp37	*flooding* from *sanitary sewers*	*(No./100 km sewer/year)*	*L1*
	wOp38	*flooding* from combined sewers	*(No./100 km sewer/year)*	*L1*
	wOp39	*surface flooding*	*(No./100 km sewer/year)*	*L1*
	wOp40	Sewer *collapses*	*(No./100 km sewer/year)*	*L2*
	wOp41	Pump failures	*(hours/pump/year)*	*L2*
	wOp42	Power failures	*(hours/pumping station/year)*	*L2*
CSO CONTROL	*wOp43*	*CSO* control	*(%)*	*L3*

SUBGROUP	Code	Indicator	(unit)	Level
WASTEWATER AND SLUDGE QUALITY MONITORING	*wOp44*	Wastewater quality tests carried out	(-/year)	L1
	wOp45	- BOD tests	(-/year)	L3
	wOp46	- COD tests	(-/year)	L3
	wOp47	- TSS tests	(-/year)	L3
	wOp48	- total phosphorus tests	(-/year)	L3
	wOp49	- nitrogen tests	(-/year)	L3
	wOp50	- *faecal E.coli* tests	(-/year)	L3
	wOp51	- other tests	(-/year)	L3
	wOp52	*sludge* tests carried out	(-/year)	L2
	wOp53	Industrial discharges tests carried out	(-)	L2
VEHICLE AVAILABILITY	*wOp54*	Vehicle availability	(No./100 km sewer)	L3
SAFETY EQUIPMENT	*wOp55*	Gas detectors	(No./employee)	L3
	wOp56	Permanently installed gas detectors	(%)	L3

QUALITY OF SERVICE INDICATORS (wQS)

SUBGROUP	Code	Indicator	(unit)	Level
POPULATION SERVED	*wQS1*	Resident population connected to *sewer system*	(%)	L1
	wQS2	Resident population served by WWTP	(%)	L1
	wQS3	Resident population served by *on-site systems*	(%)	L2
	wQS4	Resident population not served	(%)	L2
TREATED WASTEWATER	*wQS5*	Treated wastewater in WWTP	(%)	L1
	wQS6	- *preliminary treatment*	(%)	L3
	wQS7	- *primary treatment primary*	(%)	L3
	wQS8	- *secondary treatment*	(%)	L3
	wQS9	- *tertiary* treatment	(%)	L3
FLOODING	*wQS10*	*flooding* of properties from *sanitary sewers* in *dry weather*	(n.º/1000 properties/year)	L1
	wQS11	*flooding* of properties from *sanitary sewers* in *wet weather*	(n.º/1000 properties/year)	L2
	wQS12	*flooding* of properties from combined sewers in *dry weather*	(n.º/1000 properties/year)	L2
	wQS13	*flooding* of properties from combined sewers in *wet weather*	(n.º/1000 properties/year)	L2
	wQS14	*surface flooding* of properties in *wet weather*	(n.º/1000 properties/year)	L1
INTERRUPTIONS	*wQS15*	Interruption of wastewater collection and transport services	(%)	L2
REPLY TO CUSTOMER REQUESTS	*wQS16*	New *service connections* efficiency	(days/new connection)	L3
	wQS17	*service connection repair* time	(days/repaired connection)	L3
	wQS18	Average response time to empty *septic tanks* or pits	(days/request)	L3
COMPLAINTS	*wQS19*	Total *complaints*	(No. /1000 inhab./year)	L1
	wQS20	- *blockage complaints*	(No. /1000 inhab./year)	L2
	wQS21	- *flooding complaints*	(No. /1000 inhab./year.)	L2
	wQS22	- pollution incidents *complaints*	(No. /1000 inhab./year)	L2
	wQS23	- odour *complaints*	(No. /1000 inhab./year)	L2
	wQS24	- rodents related *complaint*s	(No. /1000 inhab./year.)	L3
	wQS25	- customer account related *complaints*	(No. /1000 inhab./year)	L2
	wQS26	- other *complaints*	(No. /1000 inhab./year)	L3
	wQS27	Response to *complaints*	(%)	L2
	wQS28	Responsibility for third party damages	(%)	L2
IMPACTS ON TRAFFIC	*wQS29*	Traffic disturbances	(km/ traffic interruption)	L3

ECONOMIC AND FINANCIAL INDICATORS (wFi)

SUBGROUP	Code	Indicator	(unit)	Level
ANNUAL REVENUE	*wFi1*	Unit revenue	(US$/p.e./year)	L1
	wFi2	- service revenues	(%)	L2
	wFi3	- other revenues	(%)	L3
	wFi4	Industrial revenues	(%)	L2
ANNUAL COSTS	*wFi5*	Unit total cost per p.e. [11]	(US$/p.e./year)	L1
	wFi6	Unit total cost per length of sewer [11]	(US$/km sewer/year)	L1
	wFi7	- unit running cost per p.e. [12]	(US$/p.e./year)	L2
	wFi8	- unit running cost per length of sewer [12]	(US$/km sewer/year)	L2
	wFi9	- unit capital cost per p.e. [13]	(US$/p.e./year)	L2
	wFi10	- unit capital cost per length of sewer [13]	(US$/km sewer/year)	L2
COMPOSITION OF ANNUAL RUNNING COSTS PER TYPE OF COSTS	*wFi11*	Internal manpower costs	(%)	L3
	wFi12	External services costs	(%)	L3
	wFi13	Energy costs	(%)	L3
	wFi14	Materials, chemicals and other consumable costs	(%)	L3
	wFi15	Other running costs	(%)	L3
COMPOSITION OF ANNUAL RUNNING COSTS PER MAIN FUNCTION (internal and outsourced)	*wFi16*	General management functions costs	(%)	L3
	wFi17	Human resources management functions costs	(%)	L3
	wFi18	Financial and commercial management functions costs	(%)	L3
	wFi19	Customer service management costs	(%)	L3
	wFi20	Technical services functions costs	(%)	L3
COMPOSITION OF ANNUAL RUNNING COSTS PER TECHNICAL FUNCTION ACTIVITY	*wFi21*	Wastewater treatment running costs	(%)	L3
	wFi22	*sewer system* running costs	(%)	L3
	wFi23	Wastewater quality monitoring running costs	(%)	L3
	wFi24	Support services running costs	(%)	L3
COMPOSITION OF ANNUAL CAPITAL COSTS	*wFi25*	Depreciation costs	(%)	L3
	wFi26	Net interest costs	(%)	L3
ANNUAL INVESTMENT	*wFi27*	Unit investment	(US$/p.e.)	L2
	wFi28	- investments for new assets and expansion of existing asset capacity	(%)	L3
	wFi29	- investments for asset *refurbishment* or *replacement*	(%)	L3
EFFICIENCY INDICATORS	*wFi30*	Total cost coverage ratio	(-)	L1
	wFi31	Operating cost coverage ratio	(-)	L1
	wFi32	Delay in accounts receivable	(days equivalent)	L3
	wFi33	Investment ratio	(-)	L2
	wFi34	Contribution of internal sources to investment (CTI)	(%)	L1
	wFi35	Average age of tangible assets	(%)	L2
	wFi36	Average depreciation ratio	(-)	L3
	wFi37	Late payments ratio	(-)	L2
	wFi38	Inventory value	(-)	L3
LEVERAGE INDICATORS	*wFi39*	Debt service coverage ratio = DSC	(%)	L2
	wFi40	Debt equity ratio	(-)	L2
LIQUIDITY INDICATOR	*wFi41*	Current ratio	(-)	L1
PROFITABILITY INDICATORS	*wFi42*	Return on net fixed assets	(%)	L2
	wFi43	Return on equity	(%)	L1
	wFi44	Return on capital employed	(%)	L2
	wFi45	Asset turnover ratio	(-)	L2

11 Indicators *wFi5* and *wFi6* shall be used in alternative. *wFi6* applies if these costs are referred just to sewer systems.

12 Indicators *wFi7* and *wFi8* shall be used in alternative. *wFi8* applies if these costs are referred just to sewer systems.

13 Indicators *wFi9* and *wFi10* shall be used in alternative. *wFi10* applies if these costs are referred just to sewer systems.

Appendix 4 - SIGMA Lite WW

INTRODUCTION TO SIGMA LITE WW SOFTWARE

This Appendix introduces the SIGMA Lite software, provided on the CD-ROM that accompanies this manual.

Sigma Lite WW is a freeware package prepared by Instituto Tecnológico del Agua (ITA)[14]. It has been developed similarly to the earlier equivalent software provided with the manual for water supply performance indicators. SIGMA Lite is a simplified version of SIGMA Pro, which is professional software for performance indicator evaluation. The Pro version has now been developed for water supply PIs. SIGMA Lite WW contains the set of performance indicators for wastewater services given in this manual.

SIGMA Lite assists the user in implementing the wastewater PI system given in the manual. It should be used once selection of the set of PIs to be evaluated has been made (Section 5). The main features of the SIGMA Lite software are:

- Incorporation of the complete set of performance indicators from the IWA proposal as a stand alone PI evaluation system;
- Easy to operate, with automatic calculation of PIs from input variables - thus reducing risk of calculation errors;
- Facility to export the results to an MS-Excel® spreadsheet file for further interpretation and processing.

SIGMA Lite versions of the SIGMA software are prototype tools designed to evaluate the potential of the IWA performance indicator systems. The Lite version is distributed with all copies of the IWA's Manuals of Best Practice on Performance Indicators for Water Supply Services (Alegre *et al.*, 2000) and for Wastewater Services (this manual).

The professional version of SIGMA, developed as an independent project by ITA, has more features than the SIGMA Lite version, for instance: customizable PI System; benchmarking initiative support; possibility of working with several files simultaneously; customisable graphs; reminder of data inputs; security access management; networking; restricted file use for an individual undertaking; historical analysis; customisable reports with graphics; and multiple undertaking management.

The CD content that accompanied this book can now be found at
http://www.sigmalite.com/iwamanual/

SIGMA LITE WW DESCRIPTION AND PROCEDURES

MENUS AND COMMANDS

SIGMA Lite has a friendly graphic user interface and management of the indicators and variables. In Figure 10 a brief description of all menus and commands available in the software is presented.

14 Universidad Politécnica de Valencia, Spain (http://www.ita.upv.es).

Table 17. Menus and commands in SIGMA Lite.

Menu	Command	Description
File	Save *	Last data inputs (variables and context information), last PI calculation and also the options set for the program are saved with the software and will be kept even if you close SIGMA Lite.
	Printer setup...	Allows the user to choose a printer and to setup some printing options (e.g. paper size, paper source, portrait or landscape orientation).
	Page setup...	With this command, the user can configure the page where a PI assessment report spreadsheet is printed: header, footer, margins, position in the page, scale and other options.
	Print...	Before printing, it is possible to select the PI and variable groups for the report.
	Exit	Closes SIGMA Lite
Edit	Undo *	Cancels last actions performed in reverse order.
	Cut *	Removes a selection.
	Copy *	Copies a selection.
	Paste *	Places cut or copied selection at an insertion point.
	Delete	Clear a selection.
	Select all	Selects everything in the current window.
	Cancel selection	Deselects the selection made in the current window.
Select	PI selection...	Opens a window where the user can choose the performance indicators to be assessed.
Data	Show PI *	Opens a tree view of the selected PI
	Show variables *	Opens a tree view of all the variables involved in the assessment of selected PI
	Enter all variables	According to user preferences, this command opens a window with variable cards or a spreadsheet where variable values and confidence grades can be entered.
	Enter selected variables	This command unlocks the variable values and confidence grades editor. After selecting it, the padlock icon will appear open.
	Edit context information	Context information data can be entered or retrieved anytime by using this command to open the window.
	Evaluate *	Performs the calculation of selected PI with the input data.
Tools	Show spreadsheet *	Opens the window containing not only the results obtained by SIGMA Lite for the selected PI but also the variables and context information data inputs.
	Print spreadsheet *	Send the spreadsheet page to the default printer.
	Export spreadsheet *	Allows the user to save data shown in the PI, variables and CI spreadsheets in other format files that can be opened and processed in other programs. Available formats are: Microsoft Excel Workbook (.xls), HTML Data (.html) and Tabbed text (.txt)
	Options *	This allows the user to customize SIGMA Lite running: Explorer options – use one or multiple windows Start options – check for new versions at startup and show maximized options Visualization options – status bar, clock and date, PI and variables tree views, tips on record cards, windows positions saving Processing options – keep previous values, ask for or replace a previous value when calculating variables Data options – variable entering mode (record cards or spreadsheet) and decimals number definition
	Customize	Toolbar and tool buttons selection.
Window	Cascade	Adjusts the position of windows in such a way that all title bars remain visible.
	Horizontal tile	Shows all windows simultaneously so that they fill the screen horizontally but do not cover each other.
	Vertical tile	Shows all windows simultaneously so that they fill the screen vertically but do not cover each other.
	Close window	Closes active window.
	Show PI *	Similar to "Show PI" in the Data menu.
	Show variables *	Similar to "Show variables" in the Data menu.
Help	Contents, Help Index and Search	Open a window containing two tabs, one for reading SIGMA Lite help and the other for searching keywords.
	About SIGMA *	Information on SIGMA Lite version, copyright and authors.

* This command has equivalent tool button on the toolbar.

INSTALLING SIGMA LITE

Installing SIGMA Lite is an easy task. The user may find the necessary file in the CD-ROM provided with this manual or download it from the SIGMA website (http://www.sigmalite.com).

The installation of SIGMA is self-explanatory, and the user only has to follow the instructions provided by the installation wizard.

The most recent version of SIGMA Lite software is available free of charge in the SIGMA website. If the software is already installed, the check for new updates is automatic if the box "Check for new versions at start-up" in the Start-up options is ticked. Also the user may compare the version installed with the latest version available by clicking the button "Check now" when the computer is linked to the Internet.

GETTING STARTED WITH SIGMA LITE AND RECOMMENDED PROCEDURE FOR PI ASSESSMENT

After a brief description of SIGMA Lite, some guidelines are provided for the first use of this software.

The first time SIGMA Lite is used, the application wizard allows the user to configure the program for use within a particular undertaking. The main wizard questions relate to:

- undertaking identification – name, reference date for the values of the indicators and the responsibility for the data
- context information – the wizard presents a form, where each tab corresponds to a CI group, according to undertaking, system and region profiles; the user can edit context information later, or at any time, with the option "Edit context information" in the Data menu. Although context information values have no effect on final values of the indicators, when performing benchmarking analysis with other undertakings, CI values should be included as explanatory factors for indicator values.
- selection of performance indicators – the first time SIGMA Lite is used, a window for PI selection is presented. The user may calculate as many PI as applicable from the list proposed by this manual, according to the significance of the individual PI and also to data availability. This selection may be changed any time with the command "PI selection...".

After this third step the wizard closes and the user can start entering data for the selected PI assessment, using the Data menu. When SIGMA Lite is started again, the wizard will not appear.

The recommended procedure for PI evaluation using SIGMA Lite is shown in Figure 10, where each step is followed by a possible command. After selecting a set of PIs from this manual for assessing undertaking performance, the SIGMA Lite user must follow steps 1 to 4 in order to process data and report results.

The CD content that accompanied this book can now be found at
http://www.sigmalite.com/iwamanual/

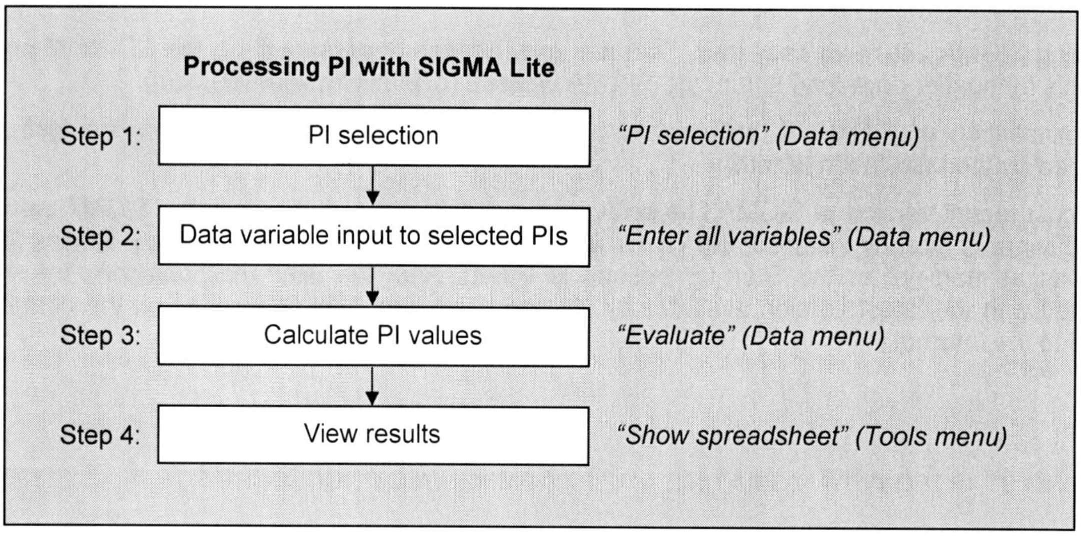

Figure 10. Recommended PI evaluation procedure using SIGMA Lite.

First, the PIs of interest are selected in the software (Step 1), which automatically accesses the variables involved in the PI processing. The user may then input the data values (Step 2) and calculate PIs (Step 3). The results are presented in a spreadsheet (Step 4) that can be exported to a MS-Excel® file for further processing and analysis, such as graph building. SIGMA Lite only stores the last data entered.

There are alternatives for the commands referred to in Figure 10. For example, an alternative procedure for entering data - Step 2 – is achieved by double clicking variables on the tree view (Show variables, Data menu) and browsing variable cards using the navigation buttons. A similar procedure may be adopted for checking PI results - Step 4 (Show PI, Data menu).

Finally, it is recommended that users read SIGMA Lite Help (Help Menu or F1) for further information and tips on the use of the software.

BIBLIOGRAPHY

Alegre, H. (2002) **Performance indicators as a management support tool**. *in* Mays, L.W., **Urban Water Supply Handbook**, MacGraw-Hill, ISBN 0-07-137160-5.

Cabrera, E., Dolz, R., Trull, O. **SIGMA Lite Help**, *in* SIGMA Lite Software, Version 1.5. Retrieved January, 2003, from SIGMA Lite software Web site: www.sigmalite.com

Alegre, H., Hirner, W., Baptista, J.M. & Parena, R. (2000). **Performance indicators for water supply services**. Manual of Best Practice Series, IWA Publishing, London, ISBN 1 900222 27 2 (160 pp.).